ATTENTION

ATTENTION
Beyond Mindfulness

GAY WATSON

REAKTION BOOKS

Published by Reaktion Books Ltd
Unit 32, Waterside
44–48 Wharf Road
London N1 7UX, UK

www.reaktionbooks.co.uk

First published 2017
Copyright © Gay Watson 2017

Printed and bound in Great Britain by
TJ Intenational, Padstow, Cornwall

A catalogue record for this book is available from the British Library

ISBN 978 1 78023 745 9

Contents

The Zen Master Ikkyu was once asked to write a distillation of the highest wisdom. He wrote one only word: *Attention*. The visitor was displeased. 'Is that all?' So Ikkyu obliged him. Two words now. *Attention. Attention.*

I

Attending to Attention

Tell me to what you pay attention and I will tell you who you are.

ORTEGA Y GASSET[1]

Each of us literally chooses, by his way of attending to things, what
sort of a universe he shall appear to himself to inhabit.

WILLIAM JAMES[2]

Pay attention

We hear the command to 'Pay attention!', and the years drop
away as we are returned to our desks at school. From childhood we have been exhorted to pay attention and yet in truth very
little attention is ever paid to attention itself. What we do pay
attention to is most generally either its absence or the object or
content of the attention that is required, ignoring the process itself.
Yet attending, that overlooked process, is at the very heart of who
we are, and how we become that person we think we know. As a
Buddhist scholar wrote: 'It is as if the accomplishment of mere tasks
is of primary value, while the quality of awareness with which these
tasks are undertaken is irrelevant.'[3] However, if we voluntarily and
consciously turn our awareness towards the process and not the
product, we may discover a whole new way of seeing and being.

Attention is at the heart of everything we do and think, yet it is
usually invisible, transparent, lost behind our fixation with content.
We pay attention to every moment or we let our attention wander,
but we rarely give attention to the process of attending and distraction. It is typically viewed instrumentally, in terms of what it
can achieve, and so its process and practice are overlooked. Yet

9

such sought-after traits as mindfulness and grit are founded upon attention. With the exception of the novelist Aldous Huxley, who began his 1962 novel *Island* with the word 'Attention' spoken by mynah birds trained to call 'attention' and 'here and now' as constant reminders of our forgetfulness, we pay little attention to attention. Why has this vital part of our experience been so ignored?

Here and now would seem a good time to turn attention to attention itself. Some aspects of attentional practice have become common topics of current discussion and research, both as threat, demonstrated by new media, distraction and attention deficit disorder, and as enhancement in the ever-burgeoning field of mindfulness practices. But to consider attention meaningfully, we need to consider both the *how* and the *what*: *how* our attention actually works in daily life, and to *what* we can and possibly should allot the finite resource of our attention in order to live better. We find that the processes of attention involve both those that are involuntary and some that are voluntary, under our control and will. Practices of attention are obviously targeted at the latter and will form the major subject-matter of this book. As the cultural attacks on our involuntary attention increase exponentially from the constant overload of advertising, and commercial inroads on our limited attentional capacity, so a wise cultivation of good habits of attention becomes increasingly important.

Neuroscience has revealed that the processes of neuroplasticity – the ability of the brain to alter its very structure, the pathways and patterns of the firing of neurons, in response to repeated experience – continues throughout life. Mechanisms of attention are found to be at the centre of this process. Voluntary attention, it now appears, is a skill, something that we can learn through practice, sculpting our brain patterns, enhancing the mind in ways that will deeply influence the people we become. Just as 'pay attention' has become a mere phrase rather than a conscious action, so the belief that we are 'creatures of habit' has become an unconsidered adage. Yet if, as research is showing, we are literally formed by our habits, our brains sculpted by our repeated routines, it surely behoves us to choose carefully what habits we espouse. Practice denotes conscious attention; habit is often unconscious practice become so habitual that it is no longer noticed. Good habits require continual questioning.

The value of practices of attention was initially revealed to me as I wrote a book entitled A Philosophy of Emptiness.[4] This explored ideas of emptiness through time and space, from their origins in the East with Taoist and Buddhist teachings and entrance into the West through Greek philosophers, to a long period of forgetfulness under Christian supremacy, until they resurfaced in the modern world. So, a very brief diversion to illuminate ideas of emptiness is called for. For this emptiness, despite its name, does not denote mere absence or non-existence. In Taoist writings Presence, the empirical world of the ten thousand things in constant transformation, arises from the generative foundation of Absence or emptiness. Tao, or Way, can then be understood as the process of emptiness continuously giving rise to the realm of Presence, of form emerging from formlessness.[5] Taoist texts tend to be both more poetic and more concerned with foundation and generation than Buddhist texts, which are much more plentiful and more psychological, and present a far greater, more academic and worked-through philosophy of emptiness developed over centuries of study and meditative experience. These philosophical ideas of emptiness are concerned with an emptiness of essence, of form, fixity, permanence, intrinsic identity and definition. The other face of emptiness then becomes interdependence, a celebration of plurality, change and contingency. A full understanding of emptiness, which entails an appreciation of the impermanence and interdependence of everything that is far from being an absence, changes and enriches one's perspective on life and the self. What I discovered was that such an understanding required a turning away from our Western, and mostly unconscious, fixation on presence and substance. In Eastern traditions, the rigid opposition between presence and absence, existence and non-existence is more blurred and ambiguous. This gives rise to, and indeed arises from, such concepts as *sunyata*, the Sanskrit term that is usually translated as emptiness, though, as I hope the description above has indicated, it has far more positive connotations than the English word commonly embraces. An interesting example of this is that *sunya*, or empty, was the word first used in India to translate the concept of mathematical zero, so indispensable for arithmetical operations and today forming half of the binary language of modern computers.[6]

The West has traditionally resisted change and impermanence and followed the ways of presence and substance, embracing a world that obeys abstract laws and an immortal soul that may allow us to escape death, and a god or form that sits beyond or above ever-changing reality, which, since it is unchanging, is somehow more real. For those brought up in the usually unquestioned Western outlook, appreciation of emptiness, impermanence and interdependence asks for a kind of figure/background reversal, away from this historical and now-unconsidered concentration on substance and presence. Such a reversal allows us to appreciate the dynamic dependence of form on space, and of sound on silence. The picture gains depth, context and richness. Seen in this light, one can understand that Western philosophy needed deconstruction in order to loosen the hold of form and fixity, substance and presence. Such a reversal, to be followed by reconstruction of a truer balance, asks for a practice of attention. It is close attention that reveals bias, and which then allows for recalibration and balance. But such attention also requires *practice*, and what my researches into emptiness have also revealed is that practices of attention, once inextricably linked to philosophy (etymologically the love of wisdom), despite attention's importance as theoretically revealed by the neurosciences, have long been divorced from both science and philosophy. However, at the same time, in art practice, in psychology and in the contemporary popularity of mindfulness in every aspect of life, such practices are still around, perhaps even more needed in an ever-changing world where compensatory emotional seeking for certainty is so strong. Thus an exploration of the habits and practices of attention seems to follow on seamlessly from that of philosophies of emptiness.

As the Buddha asserted: 'Whatever one frequently thinks and ponders upon, that will become the inclination of one's mind.'[7] Indeed, at the beginning of the twentieth century, philosopher and psychologist William James described living creatures as 'bundles of habits', labelling those that are innate as instincts while suggesting that others, due to education, might be called acts of reason. Long before the era of neuroscience, he even described the plasticity of the brain as 'the way currents pouring in from the sense-organs makes with extreme facility paths which do not easily disappear'.[8]

More recently, composer Philip Glass has written of how he trained himself in 'the habit of attention'. He describes how he set aside each day the hours from ten in the morning until one for composition, forcing himself to sit at the piano whether or not idea or inspiration came. 'The first week was painful – brutal, actually. At first I did nothing at all during those three hours. I sat like an idiot without any idea of what to do.' Then slowly things began to change: 'I started writing music, just to have something to do.' After a few weeks: 'I found the transition from near madness and frustration giving way to something resembling attention . . . From then on, the habit of attention became available to me, and that brought a real order to my life.'[9]

We can train habits and we can lose them. In the 1980s scientist Gerald Edelman coined the term 'neural Darwinism', explaining how experience strengthens or prunes away the pathways of our neuronal firings. Each of us is thus deeply individual, the results of our individual experiences. Oliver Sacks has described how in the development of motor skills in babies and rehabilitation after injury the paths of learning differ. Everyone must discover or create his own motor and perceptual patterns, his own solutions to the challenges that face him . . . neural Darwinism implies that we are destined, whether we wish it or not, to a life of particularity and self-development, to make our own individual paths through life.[10]

Practices of attention then become practices of self-creation. They may enhance, enrich or impoverish our own lives and indeed the lives of those around us.

It is vital, then, that we discover what practices may help the development of a healthy mind and what may harm it. At this time of rapid social and technological change, as the use of computers, electronic devices, social media and texting are changing our practice and our experience, attention to attending becomes even more crucial. While much of the concern around these new habits is critical, a recent article by a Buddhist writer goes so far as to suggest that the Internet might be seen as a collective mind and that media sites that allow millions to share what one person has recorded might be regarded as supporting what he calls 'an emerging form of global meditation'.

> Witnessing an atrocity, observing injustice in action, or otherwise directly encountering the things that have historically been invisible is a way of shining the light of awareness into the dark corners of our world – much as meditation shines a light into the unexamined shadows of our mind. But the collective challenge is as daunting as the individual one: how do we bring patience, kindness, and equanimity to what we see instead of having it trigger and release the reservoirs of anger and hatred lurking within that are so ready and eager to erupt?[11]

It all depends on how we attend; the ethical ingredient of attention.

Day by day, research in the neurosciences is illuminating mental processes more clearly. Many recent findings, such as the discovery of the default mode of the brain, of the distinction between narrative and present-focused modes of attention and of the potential unhappiness of the wandering mind, open up the field in ways that are not only fascinating but central for well-being.[12] Neuroscience, however, will never provide the whole picture of our experience; we are also embodied creatures, embedded in culture, as the excerpt above shows. Psychologist Paul Broks, exploring the work of Nicholas Humphrey, suggests that 'getting one's head around the problem of consciousness, experiencing the truth of a scientific or philosophic theory, may be as much the concern of art as science.' He argues that even if the problem may one day find a solution in scientific terms, these terms may not fit into the frame of human imagination, and that artistic methods and media may prove more valuable for exploring the nature of phenomenal existence.[13]

Aside from science, other fields, ancient practices recently modernized, psychotherapy and, as Broks suggests, the various disciplines of art and craft, have long experience in practices of heightened attention. My intention is to engage with the experience of experts in all of these areas to attempt to present an exploration of attention in action, an investigation that is wide-ranging rather than deep, and experientially rather than theoretically based. I want to converse with those who are experts in attention in various fields, to inquire into the place of attention in their experience and to explore what practices have honed and continue to inform and enrich their attentional skills. A recent scholarly book on consciousness and attention states that attention should be defined functionally, whereas consciousness is defined in terms of its phenomenal character without a clear purpose.[14] I was pleased to read this, as it is my intention not to address the subject in a scholarly or descriptive manner but to explore it, functioning in different fields – attention in action, as it were. The more one thinks about attention, the more one realizes that it is at the very heart of all that we do and think. It is occurring at every moment of our daily lives. Even our nightly dreams are influenced by the residues of the day's attentional movements. A book on attention could ramify unceasingly and end up as wide as the world and as unstable as the self. I shall attempt to keep my focus on practices of attention.

Three distinct threads will weave through this exploration: first that of attention as a skill that can be learned and practised and, following on from this, a consideration on what practices may be healthy or harmful, opening up the intention behind our practices of attention. The third significant thread is a reflection on enhanced or refined attention as of importance to a secular or immanent transcendence. In our material age there is still a yearning for some form of transcendent or 'spiritual' experience. It would seem that a heightened form of experience, Freud's 'oceanic feeling', a sense of oneness with nature or the expansion that art may induce in us, may play this kind of role and are also products of a different kind of attention.

BUT TO RETURN TO ATTENTION ITSELF. Dictionary definitions should inform us as to its importance and reach.

at·tend
v. at·tend·ed, at·tend·ing, at·tends

    v.tr.
1. To be present at: *attended class.*
2. To accompany as a circumstance or follow as a result: *The speech was attended by wild applause.*
3a. To accompany or wait upon as a companion or servant.
  b. To take care of (a sick person, for example).
4. To take charge of: *They attended our affairs during our absence.*
5. To listen to; heed: *attended my every word.*
6. Archaic To wait for; expect.

v.intr.

1. To be present.
2. To take care; give attention: *We'll attend to that problem later.*
3. To apply or direct oneself: *attended to their business.*
4. To pay attention: *attended disinterestedly to the debate.*
5. To remain ready to serve; wait.
6. Obsolete To delay or wait.

From Latin: ad + tendere = to stretch to
To direct the mind or energies to; to watch over, to wait for, to expect;
To listen, to regard, consider
To look after.

To direct one's care to, to tend.[15]

The scope and gravity of these definitions should alert us to the centrality of attention to our experience. Let's take a look at them. Firstly, *to be present.* Surely this is fundamental to all experience? How can we experience at all if we are not present? And yet, how often are we truly, consciously, present here in this moment, at one with what we are experiencing? This failure is probably particularly prevalent in our contemporary, busy, urban and multimedia world, when our senses are scattered, our eyes bombarded with incoming messages, our ears surrounded by almost constant noise. For a moment, imagine you are a hunter; every sense is strained to take in the messages of the moment, sensing a surrounding world that may be dangerous, straining to pick up the traces of the animal you are stalking, equally aware that another may be tracking you. Maybe it is this feeling of the present that so many seek at the beginning of hunting season in a world where everyday experience is often so very different. Mindfulness, now so ubiquitously advertised for increased health and success, owes its popularity to its very rarity in the common daily experience at this point in time. Thousands of years back, our hunter forebears, steeped in the present, knowing no other alternative, could not have believed this was something that one day would have to be taught. For them attention was part of being alive; it *accompanied as a process and followed as a result* of just being alive.

To accompany, to take care of, to take charge of, to listen to, to wait for . . . all of these transitive meanings of to attend relate not only to an object but to a process, a manner of relationship with that object: a way of relating to the world in a manner that includes receptivity and *care*. Care rather than imposition. Stephen Batchelor, who has for decades been crafting an approach to Buddhism that is relevant to our time, relates how the Buddha, when asked if there was one thing that would secure benefit both in this world and what follows after death, replied, 'There is such a thing: care (apamada).'[16] Elsewhere he says that all skilful states 'are rooted and converge in care'. Batchelor admits that apamada is a difficult term to translate, commonly given as 'diligence' or 'heedfulness'. His own choice of 'care', he says, was influenced by Heidegger and his use of the German term *Sorge*. In a careful exposition of the word *Sorge*, he describes how it is grammatically a negative term, denoting not negligent or lazy, and suggests we understand it as a 'vigilant attention; allied to a heartfelt concern for the well-being of others and oneself.'[17] The intransitive meanings of the verb in the English dictionary repeat and encompass these meanings; to take care, to give attention, to remain ready, to wait, to serve.

The tendrils of meaning spread out in all directions, denoting a stance towards the world that is richer and stranger and further from our everyday understanding the more we attend to it. A clue, perhaps, is in the etymology – *to stretch to, to direct the mind or energies to, to watch over, to wait for, to expect, to listen, to regard, to consider, to look after, to direct one's care, to tend.* Surely attention is at the very heart of our experience, of how we are in the world? Or even more, a way that we might better *be* in the world. A contemporary writing of the word opens up even more resonance – @tention encompasses the idea that attention may be an address somewhere where we dwell, while also observing the tension that arises when such dwelling is not congruent with our experience, when being in attention has the quality of inattention.

One of the oldest of Buddhist texts, the *Dhammapada*, begins with the centrality of mind, which I think we could justifiably here read as attention:

> All experience is preceded by mind.
>> Led by mind,
>> Made by mind.
> Speak or act with a corrupted mind,
>> And suffering follows
> As the wagon wheel follows the hoof of the ox.[18]

Shortly thereafter the text tells us:

> Irrigators guide water;
>> Fletchers shape arrows;
>> Carpenters fashion wood;
>> Sages tame themselves.[19]
>> For the mind
> hard to control,
>> Flighty – alighting where it wishes –
> One does well to tame
>> The disciplined mind brings happiness.[20]

Millennia later, psychiatrist, scholar and writer Iain McGilchrist expresses it: 'the nature of the attention we bring to bear on the world changes what it is we find there.' And in turn what we find there influences the kind of attention we pay in future, thus 'differences of attention are not just technical, mechanical issues, but have significant human, experiential and philosophical consequences'.[21] He describes how our mind reaches 'out a hand to it [the world] (for that is what the word "attention" means, the reaching out of a hand)'.[22] Strange, then, that we pay so little attention to attention itself.

So, encouraged by McGilchrist and Zen masters, I ask the central question: 'Why is attention so important?' I hope by the end of this book, some answers to this question will be obvious. They are, of course, many and different. The philosophical response tells us that the unexamined life is life unworthy, life unlived. Science, particularly neuroscience, is daily highlighting the central-ity of processes of attention for our development and well-being. Psychology sees maternal attention as central to our development and psychotherapy attempts repair in cases of lack, stressing the importance of attention in ongoing experience, both to note our joys and to acknowledge our sorrows. In the absence of recogni-tion, there can be no voluntary emotional regulation, and these emotions and the energy they bind will come to haunt us. Increas-ingly common today are those who might be called Neurodharma therapists, who ally the latest findings from neuroscience with Buddhist awareness practices and a psychotherapeutic point of reference to enhance positive experience most strongly. Perhaps most important for all of us is the experiential answer – that with-out attention we would not sense, learn, develop or enjoy, and our lives would be, to all intents and purposes, cheapened, half-lived.

Answers to how we may best practise or hone our attention are possibly even more diverse. Beyond attention lies engagement. But attention comes first. Before we can engage wisely – in any field – we must pay attention. Paul Kingsnorth understood this deeply, writing of his first meditation retreat following years of ecological activism.

'Sit with it,' the teacher said. It is a common Zen response, and though some see it as a kind of shoulder-shrugging, to me it looks like the opposite. What it really says is: *Pay attention*. Our culture is hopeless at paying attention. It glorifies action and belittles contemplation. Responses to the ecocide currently unfurling around us are usually couched in aggressive demands for immediate 'action' – any action, it seems, however ineffective, is better than none. But it doesn't work like that . . .

Before you can act on anything with effectiveness, you have to understand it – and that is where the sitting comes in. That is where the attention matters. That is when the stripping back of your self before the indifference of nature will come to serve you.

If we sit with the earth, with the trees and the soil and the wind and the mist, and pay attention, we may know what to do and how to begin doing it, whatever burden we carry with us as we walk.[23]

How will I define attention in the following pages? Infuriatingly it will be found in many, if not all, of the meanings given above and, moreover, in its many different forms – from close focus to wide awareness, from mindfulness to mind wandering. As I have travelled through my readings and researches, and you the reader follow me through the results, you will meet the word and its practice used in different ways. For it is my intention to follow the usage of those I am engaging with and, rather than trying to come to any final answers, to awaken curiosity and shed light on to process. In short, no equations, no QED, no self-help, but attention defined experientially, as lived by a multitude of different voices: hopefully lived better through attending to attention, paying attention to attending.

In the following pages I want to explore processes of attention through various aspects of our lives. I want to talk to experts in attention; scientists who understand how these processes work, artists who understand the results and rewards of close attention, meditators, psychotherapists, body workers – all those whose work and experience encourage and allow them to pay attention to attention itself, rather than remaining in the sphere of content or object. I will attempt a phenomenological approach in the sense given by philosopher Henry Bortoft that 'phenomenology is a shift of attention within experience, which draws attention back from *what* is experienced – i.e. *where* the focus of attention is on the *what* – into the *experiencing* of what is experienced'.[24] I want to keep in touch with our experience, for each one of us is at base, however much we tend to forget this, the expert of our own attention, and the only one who can engage the intention to enhance it.

What this book will not be is a scholarly exposition of the neuroscience of attention, or a self-help book of exercises to aid in training attention. What I hope it will be is a wide-ranging discussion around many aspects of attention and the thoughts and practices of those most skilled in its use. My task will be to attempt, as seamlessly as possible, to stitch the individual testimonies into what I hope may form some kind of patchwork quilt, a serviceable whole that does not lose the variety of the different scraps of material, the particular voices that comprise it. When I used this metaphor with Jane Hirshfield, she beautifully suggested it might be a 'kaleidoscope of voices'. At the end I hope at the least that you, the reader, will close the pages with questions and an enhanced curiosity; a curiosity that will challenge habit and encourage an openness to surprise and an intention to pay attention more closely to the everyday aspects of life, with less expectation and more care.

I WOULD LIKE TO START, in Chapter Two, by going backwards in time, to reflecting on early practices of mindfulness and meditation and the theories that support such practices. Buddhist and Taoist texts present a very different slant on experience from the one with which we in the contemporary West are familiar. Yet their influence surely pervaded early Greek thought, the very foundation of Western civilization. Such influences, however, particularly those of the practices of attention, went underground for a long period of Western history. Only in the beginning of what we may call modern and postmodern times have the awareness practices, and the songs of emptiness that arose from them, been heard again above the siren calls of substance, truth, certainty and presence. Interestingly, however, the theories and the practices would seem largely to have become disconnected. While traces of understanding of contingency, the indeterminacy of existence and emptiness are evident in the work of many contemporary philosophers from Nietzsche, to Heidegger, Wittgenstein, Derrida and Merleau-Ponty, a concern with embodied practice is pretty much absent from all except the last. And while understanding of emptiness and attention is much displayed in contemporary art, music and literature, it is rarely documented as such. This chapter will also reflect on the huge popularity of the contemporary mindfulness movement in all the fields of its engagement.

After travelling through a brief history of practices of attention I will turn, in Chapter Three, to neuroscience and contemporary explanations of the ways of attention and their involvement in our experience. From such ever-increasing knowledge, new practices are arising: ways of expanding awareness, training our minds to enhance healthy experience or to evade unhelpful developments. This chapter will reflect on recent scientific discoveries and their resonance, in different language, with the beliefs and the practices of the previous chapter. It will particularly emphasize the finding that attention is a skill, and one that can be exercised and optimized. It will explore the distinction between focused attention and widespread awareness, between task orientation and what has become known as 'default mode', and consider, in the words of an influential scientific paper, whether 'a wandering mind is an unhappy mind'. It will reflect on attention as part of self-narration and creative attention that seems sometimes to transcend self-concerns. The emphasis will be on the implications of neuroscientific findings for everyday life and how we may best use this knowledge to practise healthy habits and evade harmful ones.

Turning to emotional attention in Chapter Four, I will consider the ways in which attentional practices form a child's development, with particular emphasis on attention received in the shared attention of caregiver and child, thus attending to attention both as attention-giving and attention received, and to joint attention or 'distributed cognition'. Wounds received from inadequate attention in early childhood often form the roots of what brings unhappy adults into psychotherapy where later therapeutic attention, a joint attention between therapist and client, perhaps mirrors that of early experience, and may help to heal the suffering caused by earlier deficits. Again here, recent research from the neurosciences has deepened our understanding of both child development and the earlier theories of psychotherapy; for example translating such earlier schemes as John Bowlby's Attachment Theory and indeed all talking therapies into physiological terms.

I will be particularly concerned with the use of attention in the outlook of Mindfulness Based Cognitive Therapy, now one of the most popular and governmentally authorized forms of psychotherapy. I will also be looking at those influenced by more traditionally orientated awareness practices such as Core Process Psychotherapy in the UK, and the Contemplative Psychology of Naropa University in the USA.

Following on from reflection on attention in the field of development early and late, I will turn in Chapter Five to ways of training the skill in educational fields. Subsequent chapters will turn to attention in the arts. Artists have long known the significance of attention. Much of art and its value, I believe, arises from artists drawing our attention to what we would otherwise pass by. Through imagination, strange juxtapositions, new perspectives, a fresh explanation for a taken-for-granted scenario, artists re-vision the everyday. Here, a writer, William Least Heat Moon, musing on this, I think, gets to the very heart of attention:

Sitting full in the moment I practised on the god-awful difficulty of just paying attention. It's a contention of Heat Moon's – believing as he does, any traveler who misses the journey misses about all he's going to get – that a man becomes his attentions. His observations and curiosity, they make and remake him.

Etymology: *curious*, related to *cure*, once meant 'carefully observant'. Maybe a tonic of curiosity would counter my numbing sense that life inevitably creeps toward the absurd, could provide a therapy through observation of the ordinary and obvious, a means whereby the outer eye opens an inner one. STOP, LOOK, LISTEN, the old railroad crossing signs warned. Whitman calls it 'the profound lesson of reception'.[25]

But how to keep that openness, that receptivity? How to make things new? This is the task of training and trained attention – to keep open the freshness. When we travel to unknown places, or find ourselves among unknown people, there is often that feeling of release, of excitement, of possibility – the *curiosity* of which Least Heat Moon writes. For we have to bring something to what we meet. Our path is not just laid down before us; we bring it into being even as we walk it, by the manner in which we walk it. Rather than following the path, we create our path in every moment by bringing to it our attention, imagination and creativity. It is interesting that the very word used in Sanskrit for meditation or awareness practice, *bhavana*, comes from the root meaning *being*. It denotes bringing into being and cultivating.

As we grow older, we pay less attention to our daily life. It is the unexpected, the change that grabs our notice, while much of the usual life passes by almost unconsciously. I sometimes wonder if this is not why time seems to pass more and more quickly as one ages – as more and more of what occurs is unnoticed, unattended. Neuroscience corroborates this feeling. Research has shown that once neuronal connections are laid down in the early period of development, there is a need for stability and thus less plasticity in the system. After the initial period of development, plasticity is activated only when something important, surprising or novel occurs, or if we make the effort to pay close attention.

If we consciously make that effort, attention need not always demand novelty. The novelty may be in the looking, not in the object. Wendell Berry writes of the purified attention that finds newness even in the entirely familiar, whereby the very intimacy with our surroundings, held with imagination and love (the tending the heeding, the listening of the definition?), reveals the underlying unpredictability and possibility. He says 'To know imaginatively is to know intimately, particularly, precisely, gratefully, reverently, and with affection.' And he speaks of such a loved and attentively known place:

> it is always, and not predictably, changing. It is never the same two days running, and the better one pays attention the more aware one becomes of these differences. Living and working in the place day by day, one is continuously revising one's knowledge of it, continuously being surprised by and in *or* about it. And even if the place stayed the same, one would be getting older and growing in memory and experience, and would need for that reason alone to work from revision to revision.[26]

The limits of knowledge of a place, Berry suggests, are not in that place, but in our minds. And it is in the practice of our minds, the practice of awareness, that we can struggle to better ourselves. Practices in attention are practices that encourage us to see what is there and to know it as contingent, impermanent and unique, part of an ever-changing kaleidoscope of which we too are but an element. Composer John Luther Adams also writes of place that

Over the years I've come to appreciate the different quality of experience that comes from staying in one place for an extended time . . . The longer we stay in one place, the more closely we pay attention, the deeper and richer the layers of experience we discover. Space is the distance we travel between here and there. The space we inhabit is *place*. Through patience and deep attention to where we are, we transform empty space into living place.[27]

Artists are those who have trained themselves to pay attention, to see, hear and respond most closely, and in their works encourage the rest of us, their audience, to share their vision. Thus artists, of words, image, sound and movement, will be foremost of those consulted in this exploration of embodied practices of attention.

Finally I will attempt to check us out against a map of where we may have been, and where we have arrived at the end of this particular journey, which can be only a temporary halt. The three threads behind the enquiry will hopefully be revealed as part of the pattern: attention as a skill that may be practised; some idea of what practices are indeed skilful and what unhelpful; and that third thread of an everyday sublime – an enhanced attention that may provide a sense of meaning and our place in the world. Here two poets perhaps can point to the significance I can only stumble towards. First Mark Doty, speaking of a painting, a painting of objects, a still-life.

Someone and no one. That, I think is the deepest secret of these paintings, finally, although it seems just barely in the realm of the sayable, this feeling that beneath the attachment and appurtenances, the furnishing of selfhood, what we are is attention, a quick physical presence in the world, a bright point of consciousness in a wide field from which we are not really separate.[28]

Another poet who utterly understood this relation between world and attention is Rainer Maria Rilke. He also appreciated this as our duty to world, expressed below first in a letter and secondly from the *Ninth Duino Elegy*.

> Everywhere transience is plunging into the depths of Being . . . It is our task to imprint this temporary, perishable earth into ourselves so deeply, so painfully and passionately, that its essence can rise again, 'invisibly', inside us. We are the bees of the invisible. We wildly collect the honey of the visible, to store it in the great golden hive of the invisible.[29]

> Are we, perhaps, just here for saying: House
> Bridge, Fountain, Gate, Jug, Olive tree, Window, –
> Possibly: Pillar, Tower? . . . but to say, understand, Oh to say,
> More intimately than the things themselves
> Ever thought to be.[30]

So in attention we have presence, receptivity, service, care, openness, curiosity and cultivation. Training the mind to attend to all of these – to the process of attention itself – we find our interface with world.

PART ONE *Creating Attention*

2

The Attentive Art of Meditation and Mindfulness Practices

Time flies like an arrow, so be careful not to waste energy
on trivial matters. Be attentive! Be attentive!

MASTER DAITO KOKUSHI[1]

Eastern Origins

Buddhism

Over two thousand years ago a sage, Gotama of the Sakya family, truly comprehended the importance of attention and mastered its practices, in consequence of which he became known as Buddha, the enlightened or awakened one.[2] Around his experience, his 'awakening' or enlightenment, eventually evolved what became Buddhism, the fourth largest of the world religions. There has always been some question as to whether or not Buddhism should rightly be termed a religion. It has no creed or creator god and its story of world creation is no longer literally upheld by anyone. What it does have is a profound concern with the great questions of life and death: how to live well, how to avoid suffering and how to treat others. I have always thought of Buddhist teachings as providing the 'first psychology', for Gotama analysed his own experience, his moment-to-moment consciousness, and came to some conclusions that are far from obvious but which are ascertainable by anyone through close attention and long practice. In fact he encouraged his followers to trust their own experience, and on his death exhorted them to be a lamp unto themselves.

Gotama taught that everything existent displays three marks. All phenomena are impermanent and subject to change: they are, to this extent, unsatisfactory and causing of pain, and they are what is termed not-self – that is to say they are not self-identical; they are dependent in some way upon something else. Nothing can stand alone, uncaused, independent, ultimate. At the very centre of his discoveries is what has now become known as the teaching of dependent arising, the profound interdependence among all phenomena. Everything, including our very self, is dependent, displaying the mutual dependence of parts and whole, causes and conditions and reliance on linguistic or perceptual designation.

This explains another huge and often puzzling teaching of Buddhist thought, that of emptiness. Despite the common translation of the Sanskrit word *sunya* as emptiness, this does not imply non-existence or lack, it refers to the emptiness of independence and self-identity: it points to the very inter-dependence, the co-arising of phenomena described in the previous paragraph. In Early Buddhist teachings this emptiness is largely confined to the self or rather the 'not-self'. Again this does not mean that the self is non-existent, though it does mean that there is no such 'thing' as the self. It points to the fact that rather than a completed, permanent identity, what does exist is better thought of as a process – a process of selfing or of self-creation. As we shall see later, contemporary neuroscience would agree with this picture. Unfortunately in daily life and common folk psychology, we do not realize this. We act as if we are, or have, a stable, permanent self, and from this ego-centred position conceive and arrange our world. This leads to suffering or dissatisfaction. For the world and everything in it are under the sway of those three marks of impermanence, not-self and unsatisfactoriness and thus become the cause of suffering if we misperceive them as permanent and of fixed identity. We have to open our eyes, pay attention, see that the corollary of impermanence is disappointment; we may not get what we want, we will inevitably at some point lose what we enjoy and ultimately we will all die. Which is not to say that between disappointments we will not experience joy and satisfaction, all the more so if we are freed from unsustainable expectations.

Later Mahayana Buddhist teachings extended the idea of emptiness of self to all phenomena. All phenomena are subject to change and dependence. Everything is interdependent – not non-existent, but rather wrapped up in endless interdependent process. Emptiness is perhaps like the zero in mathematics, a necessary concept for understanding and creation, *sunya* being, as noted earlier, indeed the term used by early Indian linguistics to translate zero.

Gotama came to these conclusions after much practice, austerity and mind training, the result of which was his discovery of what are known as the Four Truths, but which are better seen as Four Tasks, as described by Stephen Batchelor, who has been honing a contemporary, secular and existential approach to Buddhist teachings for many years through many books.[3] Having spent years as a monk first in the Tibetan tradition, then that of Korean Zen, Stephen has held to a consistent thread in interpreting Buddhist teachings as secular in that they still address the needs of this saeculum or age, relying on possibly the earliest texts yet restating the message of the Buddha as to how to live well and ethically in our times. These four tasks provide a road map for well-being that is as valid today as it was two and a half thousand years ago. We are asked to know that there is suffering; that the world, particularly as we conceive it, is inevitably unsatisfactory; to cease creating the causes of this suffering, the emotional cause of desire and grasping and the cognitive cause of ignorance or misperception that holds us in thrall to our erroneous view of life; to realize that there is a way to liberation from the fetters of desire, hatred and ignorance; and to cultivate a way of life that will lead to such liberation.

The eightfold path of practice that leads to liberation – right intention, right thought, speech, action, livelihood, mindfulness, concentration and wisdom – is usually divided into three divisions of Ethics, Wisdom and Attentional Practices. The path of wisdom relates to the wisdom that understands the fundamental truths of interdependence and emptiness. Inseparable from such wisdom, compassion arises: from a truly embodied understanding of these truths real morality inevitably follows; an ethics of responsibility and response-ability in a shift away from an ego-centric approach to the world to a stance of engagement and realization of participation, mutual dependence, concern and interconnection. Jay Garfield has aptly described this as 'moral phenomenology', pointing out that Mahayana Buddhist ethics, rather than relying on precepts and principles, are grounded in practices of attention that result in a shift of experience from a self-central perspective to one of interdependence and an ethic of care.[4] The third branch of the path relates to intention, mindfulness and meditation. For it was through attentional practices that Gotama made the discoveries that support his teachings of both wisdom and ethics, and it is through such practice that we may follow his lead. Awareness practices may reveal to us the reality of how we construct our experience and expose our misperceptions and our reactivity. Radical attention begins by questioning our commonplace perceptions and encouraging us to separate out the layers of experience. It enables us to filter our physical sensations and our perceptions from our feelings of attraction and repulsion and the dispositions, inclinations, habits and expectations that tend to precede them and colour them. Experiencing this, it allows us to transform ourselves through the cultivation of healthier feelings, perceptions and, above all, dispositions and habits.

Attention to the way the mind works is the very heart of the Buddha's teaching, though it may today not be at the heart of all the ways the 'religion' is practised. Models are presented of the way the mind works and how human experience comes about. In the Buddha's talks, models of human being are given in terms of *nama/rupa*, often translated as mind and body, but more accurately as name and form.

'Touch, feeling, perception, intention, attention: this is *nama* (name).'[5] *Rupa* or form refers to, yet transcends, the physical, in that it also encompasses what is visible, audible, smellable, tastable and tangible, that which Batchelor describes as the 'physical sensorium in its totality'. Interestingly consciousness is not mentioned in this first list. *Nama/rupa*, the whole psychophysical reality, is rather the necessary condition for consciousness to arise and it is considered to be inseparable from consciousness, the image often being that of two sheaves of reeds standing up, each resting upon the other. Stephen Batchelor has expressed a carefully researched, if radical, take on this matter and its implications, saying that 'Nowhere in the Suttas [the Buddha's teachings] is consciousness ever included in *nama/rupa*, nowhere, not once. Not only that but on three occasions the Buddha describes consciousness as arising from *nama/rupa*, that's why they are described as two sheaves. *Nama/rupa* and consciousness are literally interdependent . . . they are in a sort of dance, in a balance. This is enormously problematic for orthodox Buddhism because it seems to state that consciousness requires *matter/rupa*.'[6]

However, a later model of *nama/rupa* together reflects the way Buddhist teachings describe human being in a fivefold description referred to as the five aggregates or heaps (*skandha*), a list comprising form, feeling, perceptions, intentions and consciousness. As Batchelor describes it: 'That is standard Buddhist teaching, and you will find that probably in most Buddhist orthodoxies, that *nama/rupa* is understood as shorthand for body and mind. In other words they read it as a dualistic mind/body categorization, so that the body dies and consciousness continues. But what I discovered fairly recently actually was the way the word *nama/rupa* was used in the Buddha's time, at least we find it in the *Brhandaranyaka Upanisad*, is that *nama/rupa* has nothing to do with body and mind, it concerns how we identify ourselves as persons – name and form. I finally found the explicit citation in the *Brhadaranyaka Upanisad*. It basically means the plurality and diversity of things that can be recognized as having a name and having a form – how we identify things, which is what in Brahmanism one tries to transcend by returning to the undifferentiated oneness of God. So *nama/rupa* is basically the opposite of God. So when the Buddha makes *nama/rupa* and consciousness mutually dependent, he is basically taking God completely out of the picture and is looking at a purely contingent and phenomenal experience. But of course this is problematic for traditional Buddhists.'

Early Buddhism provided a very sophisticated map of mental states, one that incidentally is fascinatingly resonant, though obviously couched in entirely different language, with the findings of contemporary neuroscience. These maps or models were more fully and theoretically expanded from their first appearance in the Buddha's talks in the later literature called the *Abhidharma*. In a recent book, philosopher of mind Evan Thompson presents a commentary in terms of more contemporary understanding. He describes how models presented by both Buddhism and the early twentieth-century American psychologist and philosopher William James show that what we naturally consider a continuous flow or stream of consciousness is actually made up of mind moments. James considered the flow of such moments to be smooth, while Buddhist thinkers state that this is only so to the *untrained* mind. Buddhist practices of attention are the route to a clearer and more wholesome and ethical understanding leading to greater control and happiness.

The core insight of these models of mind moments, Thompson says, 'is that how we're aware deeply conditions what we're aware of, and that how we're aware can be ethically wholesome or unwholesome'.[7] Following *Abhidharma* texts he describes contact (touch), feeling tone, perception, attention and intention, the five components seen above in the model of *nama* as those that are always present in any mind moment. Contact refers to the relationship between a sensory or mental object, the corresponding faculty and the consciousness dependent on these two elements. Buddhist teachings refer to six senses, a mental sense being added to our usual five. They also distinguish consciousness according to its sensory or mental object. For example a moment of audio consciousness would comprise the contact of an auditory consciousness, a sound and an ear. However the fact of the other ever-present factors ensures that such contact is not without qualities. Contact is coloured by some feeling tone, whether pleasant, unpleasant or neutral. Then perception discerns the object or sensation and identifies it. There is also in intention an aspect of goal-directedness, similar to intentionality according to phenomenology. Attention is the mental factor that enables consciousness to orient towards its object. It guides and binds the other factors to the object.

Philosopher Jonardon Ganeri gives another, very interesting contemporary translation of the five *skandhas* or heaps, the Buddhist processual model of human being. His translation of form (*rupa*) is embodied reacting; feelings (*vedana*) are hedonic appraising; perception (*samjna*) becomes labelling or stereotyping; mental constructions or dispositions (*samskara*) are dispositional constructing; and consciousness (*vijnana*) is conscious attending. It is interesting to note the concordance of this model of the composition of a 'self' with a statement from a twentieth-century Western philosopher. 'Our desires and preferences are not, in general, something we just note in ourselves as alien presences. To a large extent they *are* we.'[8] It is worthy of note that all the Buddhist definitions are given in terms of verbal rather than noun constructions, retaining the sense of process so often lost. I would also point out the emphasis on attending and on habituation or stereotyping.[9]

In accordance with the Buddhist emphasis on practice and training, attention as an ever-present factor is distinguished from two other mental factors – concentration and mindfulness. Concentration refers to the ability of the mind to sustain attention and focus single-pointedly on to an object. Mindfulness is the ability to retain the object in focus from moment to moment without forgetting or wandering off from it. Such 'object-ascertaining' factors are vital to meditation and mind training and presuppose attention. The sophistication of the *Abhidharma* description of mind and mental factors may explain why it considers that the stream of consciousness flows smoothly only to the untrained mind. Buddhism is concerned with training the mind.

Meditation and Mindfulness

At the heart of mindfulness and meditation is the attempt to centre ourselves in the present – to let go of obsessing about what has already happened and worrying about what might be about to occur. It is a training in focused presence and responsiveness to the needs of the moment, best suited to enabling us to live in a world that is contingent and uncertain. Our common mode is to hide behind an ill-informed belief in identity and permanence, a magical-thinking belief that we can direct the world according to our self-centred wishes, building up that self through what we desire, increasingly today in terms of consumption, whether of material goods, knowledge or skills. Truly acknowledging that the objects and beliefs that we use to bolster our identity – the props and placemarkers we occupy in the world – are no true refuge, we may find a more flexible 'groundless ground' in an embodied awareness of what is going on in our body, feelings and mind and the world around us. The teaching most commonly referred to as the heart of meditation instruction is called the *Satipatthana Sutta*. Here the Buddha instructs his followers in the foundations of mindfulness starting with contemplation of the body, followed by contemplation of feelings, mind and the objects of mind. Each division embraces contemplation in terms of teachings; that is to say in terms of their impermanence and the four truths or tasks, freed from both grief and clinging.[10]

Today two major distinct modes of meditation are taught – focused concentration with an object *shamata*, which is usually taught as the basis for the other mode, *vipassana*, termed insight meditation, a moment-to-moment mindfulness of events as they are experienced in the present. However, as my discussion with Leigh Brasington exposed, this is a very broad distinction within which other finer distinctions nestle.

Leigh Brasington is today the leading teacher of the *jhanas* (which literally translates as meditations), a type of concentrative

meditation practice common to both Hindu and Buddhist traditions that minutely distinguishes different levels of experiences
encountered during concentrative meditation. While the Buddha
himself trained in these methods for six years without gaining
enlightenment, they have been, perhaps for this very reason, somewhat ignored in Buddhist circles, though they are experiencing
a contemporary popularity. *Jhana* practice, with its roadmap of
progression through a series of recognizable states, was traditionally followed to generate a clear mind, which can then be directed
to knowing and seeing the reality of body and mind, that is, to
insight. 'Insight practices are practices – both on and off the cushion – that aim to give us experiences of the true nature of the world
in a context such that we can understand them.'[11] Leigh likens
moving through the *jhanas* to sharpening the sword with which
to cut through the fetters of ignorance, the misplaced manner in
which we conceive of ourselves and the world; the way we defend
ourselves from impermanence by upholding our egocentric views
and building them up through desire, aversion and delusion.

Leigh also has extensive experience in other traditions of meditation and indeed also of computer programming, so my discussion
with him around the subject of attention was fascinating. I met up
with him in Bristol, where he had a week free from his peripatetic
life on the road, travelling from retreat centre to retreat centre. We
talked in a café near the railway station, which was initially quiet,
then increasingly noisy as it grew closer to lunchtime. He stated
clearly that 'attention is the core of what Buddhist training is all
about', citing the precepts, the four great efforts and the foundations of mindfulness.[12] He started by telling me: 'I know about
attention from two perspectives; one is meditation teaching, concentration meditation – the *jhanas*, and the other is as a computer
programmer. It's a different form of concentration, a different form
of attention. The difference would be that the *jhanas* require one-
pointed focused attention; computer programming is one point
after another focused attention; in other words, with concentration

meditation you pick an object and that's it; there's no variety for a given amount of time, whereas in programming I'm writing one instruction after another, I have a whole idea in mind of what I'm trying to accomplish, and I have in mind how what I'm trying to accomplish fits into the larger scope of the whole program and maybe how the program fits into the larger scope of a whole system. So although it's very focused down at the one instruction after another level, there's still the movement up and down through these levels, hierarchical movement as to what's going on, and also "I'm now about to write a loop", so there's the focus on how you write a good efficient loop as well as the focus on what the loop is supposed to accomplish. So there are a lot of different things going on, whereas if I'm practising concentration meditation there's the breathing or there's the *sukha* (happiness) of the third *jhana*, and that's it. And it just stays like that as long as I stay in that state. So that though both of them require not getting distracted – and that's how I want to translate *Samadhi*. *Samadhi* is not concentration, it's indistractability. So when you are generating *Samadhi*, you are generating the ability to not become distracted.'

It would seem from Leigh's distinction between these different forms of concentrated attention, one-pointed and task-orientated, that meditators have more precise and fine distinctions between forms of attention than the scientists, who as the next chapter will show tend to discriminate between focused attention and mind wandering. He continued to make even more precise distinctions, as he also practises in Tibetan *dzogs chen* meditation, an open awareness practice in which 'I'm trying to rest in full awareness and indistractability but nothing to do with one-pointed and nothing to do with a task or a goal or anything else – just trying to be fully present with the experience.[13] And that's a third form that I have some experience with. And they are quite distinct ways of working but practising one helps with the others because they are all about indistractability.'

It is the indistractability that is the common factor in all these methods. Leigh described to me how on a *dzogs chen* retreat he would start the day by practising the concentrative *jhanas* as 'I felt one-pointed attention would bring the indistractability – would chase away all the distractions – so that I would be better at the *dzogs chen* practice which is actually far more difficult to do in terms of not getting distracted because there is not something you can lock into. Whereas with programming there is this whole magical thing that is going to happen and the task, and with the *jhanas* there is this single thing to focus on. With *dzogs chen* there is very much just the opposite – "Don't focus on anything – just be open." So to get the indistractability happening in my mind I would go to *jhana* practice.'

We continued to discuss the way that the insight form of meditation now practised in the Insight Meditation Society centres was a form of choiceless awareness and Leigh pointed out the distinctions between these different forms, starting by saying that 'All of these are different methods of attention. Choiceless awareness is not non-distraction – distraction is fine – just simply watch what's going on. The *dzogs chen* is not non-distraction but it's non-dual – or it is non-distraction but it's not one-pointed and it is non-dual. The open awareness, hopefully you don't get lost, though I fear many simply space out, it's a lack of attention, it's back to default mode running, with hopefully an increasing facility in noting distraction. So *jhanas* would be one-pointed attention, both choiceless awareness and *dzogs chen* would not be one-pointed; choiceless awareness, done properly, would be not-one-pointed, no task and dualistic, *dzogs chen* would be not-one-pointed, no task and non-dual, and then, something like Mahasi there would be not one-pointed, dualistic and a task. And then computer programming, very much not one-pointed total non-distractibility and very task-oriented. These are all different.'[14]

I asked him if there was a similarity between Tibetan *dzogs chen* meditation and the so-called 'formless' levels of the higher *jhanas*. He replied that this was an oft-asked question, and that 'there is a similarity, particularly to the first formless *jhana* of infinite space in that there is an openness here. But there is a very definite object, so it is not non-dual. There's very much an object of space, so although it has the same openness as *dzogs chen* has, in *dzogs chen* there is not an object. Lama Surya Das said that *rigpa*, the state of clear mind, is without an artefact, so there is not an object, no effort and no distraction, whereas infinite space there is an artefact – very definitely the sense of space is what you focus on and depending how deep you are in there, there's varying amounts of effort and no distraction.[15] So it's similar but because of the lack of non-duality there it's actually quite different.'

Even in the highest-level *jhana*, designated in the relevant literature as that of neither perception nor no perception, 'there's not a distinction because there is not any thing to distinguish, whereas in *dzogs chen* there are plenty of things to distinguish but you don't. Even in that state it's far more different than the *dzogs chen*. The *dzogs chen* is open and big, the neither perception nor no perception doesn't feel big, it doesn't feel small, it feels sort of close in but you can't really say anything about it, but you can definitely say something about *rigpa*.'

Like Stephen Batchelor, Leigh goes back to the *Upanishads*, the collection of ancient Sanskrit texts that expounds the central philosophical concepts of Hinduism, to helpfully distinguish what was unique about the Buddha's message. He points to an often-quoted *sutta* where the Buddha gives a teaching to one called Bahiya, which, Leigh thinks, 'is as close as it gets to the non-dual in the *suttas*'. The usual translation of this teaching is that of 'In the seen, let there be just the seen, in the heard, let there just be the heard.' However, as he explains: 'There are no articles in Pali, so in seen there is just seen, except Pali uses past participles like English uses gerunds, so we get the true sense of it in "in the seeing let there just be seeing" and it's only verbs there, there are no objects, so it drops into a non-dual state. Bahiya was a follower of the *Brihadaranyaka Upanishad*. There is a lot in that that is helpful for understanding what the Buddha was talking about. In the *Upanishad* it says, "in seeing there is the unseen seer, in hearing, the unheard hearer, in sensing the unsensed sensor, in cognizing the uncognized cognizer. This is your *atman*." And so Bahiya has been looking for this guy and the Buddha says, "No man, in seeing, it's just seeing."'

I asked Leigh at what point he thought self-awareness was lost. 'When you get skilled at the *jhanas* by the time of the second *jhana*, you are not thinking, "I am in the second *jhana*." You're experiencing happiness and there is no "I am happy," just happiness or happying, because it's more of a verb. So the first *jhana* you lose self-awareness there but you can't be aware that you have lost self-awareness because the *piti* and *sukha* is so strong and it says in the *suttas* that there is still some background thinking.[16] It's not distracted thinking but the self can flicker in. But once you settle into the second, depending on how skilfully you are into that *jhana*, how deep it is, selfing is not really happening any more. Even at the point where you want to move to the third *jhana*, or move any *jhana*, it's more like "time to move on" rather than "I should move on" or anything like that. It's just that the awareness arises – "time to move on". So one of the most important aspects of *jhana* practice is that if you spend even twenty minutes moving through the *jhanas*, your tendency to run the self-construction network, the default mode network, has been just really reduced and the mind is much quieter.[17] Now when you come out and start taking a look at the world, in other words doing an insight practice, you are looking from a much less egocentric perspective, which is going to give you a much better view of what's going on.'

Whether this attentional training can affect our default mode of mind wandering and self-centredness in any long-term way is perhaps still an open question. When I asked Leigh about this, he thought that 'Temporarily you are running another network not default. It will come back depending on what you do afterwards. You come out of the *jhana* and jump into your car and drive to work the default is going to come back pretty quickly. You stay there doing a non-verbal meditation, even like a body scan where you are paying attention to the body, the self is not going to come back, it is just a field of awareness. Or if you are listening and paying attention to *vedana* (sensations) there is not going to be a self in there, even though I am experiencing pleasant and unpleasant, it's just pleasant and unpleasant happening – watching arising and passing, there's not going to be any self, but contemplating recollections such as "I am of the nature to grow old, sick and die" the self comes back fairly quickly but there's a certain degree of imperturbability that is mentioned in the *suttas*, that it gives you so that when you are looking at "I am going to grow old, sick and die" it's not as freaky and you can see it better, and I think part of that aspect of imperturbability is still the fact that the sense of self has been decreased and even though you start doing a very dualistic personal practice like contemplating your own death it's not the self as much as it was. There is more space around it.'

Leigh has written an article on this suggesting that mindfulness and concentration are what can replace default mode network activity, the permanent switch being Awakening.[18]

One of the fascinating aspects of attention that Leigh brought up was that of speed – how fast can you pay attention? He says that one of the variables in people's attentional ability is how fast their attention is. Another variable is the way we relate to the world: 'People relate to the world in three different ways primarily – visual, auditory or kinaesthetic. In teaching meditation I need to try and figure out a student's approach because sometimes the technique can vary based on what their dominant thing is, and the examples that I would use would vary accordingly. It is very useful if I can figure out if they are primarily visual or auditory or whatever. I had one student who was a musician and he could clearly get to access concentration but every time he got there the heating unit in the meditation hall would come on and he was hallucinating music. He had been a TM (Transcendental Meditation) practitioner so I asked him to do his old mantra in the hope that it would override the auditory channel and he would then not hallucinate music, and he eventually got to the *jhanas* that way. It was a matter of getting his attention such that it was blocking the problem and yet still useful for the solution.'

I told Leigh that one of the things I most appreciated in his book was the very grounded and physical way he expresses his teaching. He said that this bodily grounded approach has been of great use to him in dealing with problems trainees have experienced when first meeting these practices. There is often a great resistance to opening our awareness to what is actually happening in our minds. In the early days of meditation retreats in the West, there were many disasters (which will be looked at later). Leigh said that while his early training did not give him the specific tools to deal with such psychological problems, it had included forms of tactile attention, such as body scans and mindfulness of breathing which were very helpful. Later, a Somatic Experiencing training had been very useful in enabling him to get people to pay attention to their bodies to get out of distress, another instance, he felt of using directed attention to deal with serious problems.

We spoke also of the idea of enhanced attention as playing some kind of a transcendental role in an increasingly secular world. Leigh said that the word he liked was 'not transcendent but trans-send – to trans-send the usual – or trans secular.' And he thought that the way to get there is through training in attention. 'It is somehow looking at what we have got here and seeing it differently. I love Nagarjuna's "Samsara is nirvana, Nirvana is samsara." It just depends on how you look at it. Stepping out of the ordinary way of interacting.' That is the way of attention, stepping out of the ordinary through attention.

WHEN I SPOKE about meditation with another experienced practitioner, Stephen Batchelor, during a teaching visit to Gaia House in Devon, he also emphasized the centrality of embodiment in practices of attention. He has described how Gotama, asked why, as the Awakened One, he needed to practise meditation at all, answered that 'During the rains residence . . . the Teacher generally dwells in concentration through mindfulness of breathing . . . if one could say of anything: "this is a noble dwelling, this is a sacred dwelling, it is of concentration through mindfulness of breathing that one could truly say this."'[19] This awareness grounded in breath is the foundation of all the contemplative practices taught by the Buddha, it leads to a dwelling in the body that links one to the entire world. As Batchelor points out, such discipline 'involves constant vigilance that prompts one to keep returning to the felt embodiment of experience that is so easily forgotten'.[20]

When I spoke with Stephen, he said that attention itself was not something that he had given much attention to as such. He considered that 'for me attention has been swallowed up by meditation and awareness and *vipassana*. All these terms in a general sense become interchangeable, but in a specific sense, there is a Buddhist term that we translate as attention, *manasikara* or in particular, *yonisomanasikara*, but on the other hand there is not a great deal written about *manasikara*.'

Batchelor translates the Sanskrit term *yonisomanasikara* as embodied attention in contrast to the traditional translation as wise, careful or reasoned attention. This is taking into account the origin of the word from *yoni*, which is literally womb. As he explains: '[This] is completely ignored in Buddhist commentaries. It's such an incredible metaphor. It's much more radical; he [the Buddha] is basically saying that attention comes from the womb. It says that this is an attention that is grounded in the very foundations of your body that is capable of nurturing new life. There is also a wonderful passage, a phrase that occurs in one of the *suttas* where he descries *ayonisomanisakara* or disembodied attention as being eaten up by thoughts. In other words, an attention that is not *yoniso* is one in which you are being swallowed up by your own thinking processes and mental chatter, so it has very much to do with embodiment.'[21]

When I suggested that *yonisomanasikara* could perhaps be described as 'the way', or the practice, of Buddhism, he agreed, saying that 'In proper usage it is almost synonymous with mindfulness today.' He agreed that mindfulness containing continuity and memory aspects could be described as the ongoing practice of attention. 'Yes, I think that's probably the better way of putting it. Because attention, as least in theory before we get into the *Abhidharma* with the ever-present factors and so on, seems to be understood as a very core primary quality of sentience really. To say that a being is sentient means that they are capable of attending to an object and putting their mind, directed with some degree of continuity, on to it, whether it's a cat looking at a mouse, or you and I looking at a book. *Manasikara* does not have intrinsic value in the sense of being a virtue. *Yonisomanasikara* would be considered a virtue but *manasikara* itself is morally neutral. The different versions of *Abhidharma* differ as to *sati* or mindfulness. In the *Abhidharma* of Asanga, which I studied, it is considered to be morally neutral, not a virtue, just a mental activity. In the *Theravadin Abhidamma* it is a virtue.[22] I can't think of an instance in which *sati* is used in a non-virtuous sense by the Buddha, but the Pali canon is about a very specific goal; to train people to become more enlightened, so obviously the more mindful the better. This is an interesting question: is mindfulness value-free? Some insist that it is always a virtue, but this is not how Asanga understood it. Asanga gives a second set of factors after the five omnipresent factors, the five so-called object discerning factors, and mindfulness is included here along with intelligence, concentration, appreciation and aspiration which have no intrinsic value. For example, the building of a harmful object such as a nuclear bomb entails intelligence, concentration etc. I think, especially when you consider that the Tibetans translate *sati* (mindfulness) as *drenba* (memory/recollection), that it is difficult to see why you should consider that to be always virtuous. It depends on how that quality is being used, and for what purpose, that gives it a moral quality. If you frame the whole discussion pragmatically, not "what are these states of mind?"

but "what do these states of mind enable us to do?" then those sorts of issues evaporate. The only way out of the loops of discussion is to take it back to practice. How you use these things is what really matters.' Such pragmatic restating becomes important when discussing the contemporary and prevalent use (and misuse?) of mindfulness today.

Before we consider the strong contemporary interest in such practices, perhaps as an antidote to the increasing abstraction (in both sense of the word) and busyness of our lives, constantly assaulted as they are by virtual and actual information overload, I would like to make a bow to the important yet far lesser known contributions of Taoist thought, and then take a small detour from Eastern beginnings to early Western understanding of such practices. For these practices, though largely long lost, are not entirely foreign to our own culture.

Taoism

I have concentrated on Buddhist rather than Taoist teachings as they are far more fully articulated over a much longer period than those of Taoism, which tend to be both more foundational and more poetic.[23] Yet the philosophy and the practices share much. David Hinton, a poet and pre-eminent translator of Chinese texts, describes: 'The ontological structure of Way (*tao*) is replicated in the structure of consciousness, thought arising from the same generative emptiness as the ten thousand things.' Thus,

> in the depths of consciousness, Way can be experienced through the practice of meditation. You can watch the process of Way as thought burgeons forth from the emptiness and disappears back into it or you can simply dwell in that undifferentiated emptiness, the generative realm of Absence.[24]

Hinton writes that much of the text of the Tao-te-Ching can be read as describing meditative awareness presented in the form of poetry rather than discursive description. The philosophy and the practices of Taoism very largely influenced and were taken up by Chan (Chinese) and later Zen (Japanese) forms of Buddhism. A story from the Inner Chapters of Chuang Tzu illustrates a concise, embodied and very Taoist presentation of ideas of self and the practice of meditation:

[Yen Hui] May I ask about the mind's fast?
'Center your attention,' began Confucius. 'Stop listening with your ears and listen with your mind. Then stop listening with your mind and listen with your primal spirit. Hearing is limited to the ear. Mind is limited to tallying things up. But the primal spirit's empty: it's simply that which awaits things. Way is emptiness merged, and emptiness is the mind's fast.'
'Before I begin my practice,' said Yen Hui, 'I am truly Yen Hui. But once I'm in the midst of my practice I've never even begun to be Yen Hui. Can this be called emptiness?'
'Yes, that's it exactly,' replied Confucius.[25]

Taoist influences were also important in Chinese, then Japanese art forms, particularly scroll paintings and poetry. Hinton has described how Chinese poetry is imbued with understanding of what he terms a 'primal cosmology' that 'locates us in the midst of a spiritual ecology of being and nonbeing'.[26] This entails attendance to the arising of time, space and the ten thousand things burgeoning forth from generative emptiness in a perpetual process of transformation: being emerging out of non-being, a perpetual unfolding of occurrence and appearing. French Sinologist François Jullien has delineated what I find a helpful and imaginative distinction between Chinese and Western approaches both to philosophy and to art. He suggests that Chinese thought is structured by a logic of respiration rather than the logic of perception taken by Greek thought, and that such a distinction defines what is considered to constitute reality. A logic of respiration based on the rhythms of inspiration and expiration rests in continuous process through its alternation of the in breath and the out breath, and thus, rather than taking a position between presence and absence, expresses the regulating alternation of emptiness and fullness from which the process of the world flows. Existence then is seen as a proceeding forth from emptiness, the undifferentiated source. On the other hand, the logic of perception ultimately embraced by the Greeks leads to a conception of reality as an object of knowledge, a separation of seer and seen, presence and absence. For the Chinese model practices of attention, meditative practices, reveal reality; the way the ten thousand things, including thoughts, ceaselessly come into presence from emptiness and, like breath, dissolve again.

Hinton gives a wonderful description of this that I would like to quote in full:

> Here's a thought experiment: If you want to know what's fundamentally true about the world, walk out into an open field and close your eyes. Then start forgetting. Forget everything you know and believe, all the knowledge our culture has accumulated, all our assumptions about the world and ourselves. Completely empty your mind. Then open your eyes and see what you encounter. The first thing you see is this physical stuff all around you. And if you've wholly emptied your mind, it is a wondrous revelation: existence, the material universe vast and deep, everything and everywhere, when there might just as easily be nothing at all.
>
> The next thing you notice is that the empty mind perceiving that wondrous existence is not separate from existence. They are a single tissue. And if you stay there over time, looking out, you realize that existence is alive somehow. Things perpetually move and change, appear and disappear. Clouds drift. Wind rustles wildflowers and trees. Day fades into night, and night into day. Seasons come and go, one after

THE ATTENTIVE ART OF MEDITATION AND MINDFULNESS PRACTICES

another. You die. Other people are born. On and on it goes. Everything is moving all the time without pause, without beginning or end.

Finally you realize that the same thing is going on in your mind – that your mind's movements are no different than a cloud's movements or the turning seasons.

. . . When we're thinking, thinking, thinking, that's when we most intensely experience ourselves as separate from the world. Meditation lets thought fall away, and with it that sense of separation. When thinking comes to a stop, all that's left is an empty mind mirroring whatever is in front of you. Like in the field when you open your eyes.[27]

A Further Detour: The Greeks

Despite the ultimate supremacy of those Greek philosophies that espoused substance and presence (even in abstraction), which became foundational for Western thought, other traditions with echoes of Eastern ideas can yet be found within Greek philosophies and indeed practices. Early on Heraclitus said 'everything flows, nothing remains' and also noted that 'men have talked about the world without paying attention to the world or to their own minds, as if they were asleep or absent minded'.

Moreover, as in the East, Greek philosophy was not mere theory, as we know the word today: it was, to a great extent, as the title of a book that illustrates this articulates, *Philosophy as a Way of Life*. Frenchman Pierre Hadot shows how much of Greek philosophy in its day was a very lived enterprise concerned, as were the teachings of Gotama, with a way to live well and ethically. To this end it taught what the Greeks called *eudaimonia*, well-being or flourishing, *ataraxia*, which is somewhat similar to the Buddhist equanimity, and *apatheia*, the overcoming of passions. Epicurus stated:

Empty is that philosopher's argument by which no human suffering is therapeutically treated. For just as there is no use to a medical art that does not cast out the sickness of bodies, so there is no use in philosophy, unless it casts out the suffering of the soul.[28]

The overlap of thought and content of Greek and Buddhist philosophy is not my concern here. I have outlined it in an earlier book. What I want to consider here is practice. For these are practices of attention and awareness. American scholar Thomas McEvilley in his monumental study *The Shape of Ancient Thought* found similarities between Early Buddhism, Epicureanism, Stoicism and Pyrrhonian Scepticism in terms of naturalistic theories of knowledge based on sense experience, naturalistic doctrines of causality, a fundamentally naturalistic ethics based on pleasure and pain rather than on absolute good or evil, and a naturalistic psychology which is the base for strategies of what he called 'interrupting mental process'. This refers to their advocacy for mindfulness that may delink or defer the processes of perception and action in order to foster a considered response to a stimulus rather than a hasty and habitual reaction. As with Buddhist trainings, the aim is to educate the mind, to gain self-mastery, and to bring one's own mind, over which one may have, with practice, some control, into alignment with the flow of events, the impermanence of the world, over which one has no control.

As Hadot describes it: 'Attention (*prosoche*) is the fundamental Stoic spiritual attitude.'[29] For both Stoics and Epicureans, he says, philosophizing 'was a continuous act, permanent and identical to life itself', and for both schools 'this act could be defined as an orientation of the attention'.[30] Attention to the present moment is the key. The path to the quietude sought by the Greeks, the path to the elimination of concepts that foster linguistic reification, involves a direct relationship with the present moment. Thus Greek teachings are similar to mindfulness in their emphasis on a non-conceptual, non-judgemental awareness of the present moment. If no traditions of what we would call meditation have come down to us from the Greeks, there most certainly did exist practices of imagination and thought experiments. According to Hadot, meditation in their case was 'the exercise of reason'. One important aspect of this, that Hadot emphasizes, is the cosmic dimension of such spiritual practices. The goal is to go beyond the self, to align the self with universal or cosmic reason. The feeling of connection with the whole, being a part of nature and a portion of universal reason, is an essential element of Greek practice, and one that transforms the feeling one has of oneself.[31] Today, we might well see this as an antidote to the limited individualism of our own age.

Unlike Buddhist practice, Greek spiritual training has left no somatic form 'but is a purely rational, imaginative or intuitive exercise that can take extremely varied forms'.[32] Such philosophy as a way of life, however, began to die out in the West with the coming of Christianity and the emphasis on faith. Practices remained in the monastery, and theory, which Hadot describes now as philosophical discourse, divorced from practice and everyday life, took place in academies, somewhat different academies from those of Plato. Thus began the separation of theory and practice, of reason and belief, of mind and body. Although many of these splits have been addressed in more recent philosophy, I think one can say that there is still a gulf between theory and practice. While much has been written about engagement with practice, it is still quite difficult to find those that actually engage with practice. Eugene Gendlin, the creator of an embodied practice called Focusing, is far better known in the world of psychotherapy than he is in philosophy. However, there exists also a current revitalization of interest in ancient practices. Contemporary German philosopher Peter Sloterdijk has said: 'The emancipation of practice from the compulsive structures of Old European asceticism . . . may possibly constitute the most important intellectual-historical and body-historical event of the twentieth century.'[33]

As well as the contemporary appeal of mindfulness in all forms of life that we will consider next, there would appear to be also a resurrection of interest in Greek, particularly Stoic, thought. It began some decades ago with the appearance of small pocket-sized editions of the thoughts of Marcus Aurelius, and today has grown into the popularity of Stoic weeks and Stoic camps. There exists a *Stoicism Today* team comprised of academics based at the University of Exeter, Kings College and Queen Mary University of London and psychotherapists working in the UK and in Canada to create 'Stoic resources for the modern day'. There have to date been five annual Stoic Weeks. In 2016 the Stoic Week handbook contained readings, links to audio, video and group discussions and daily practical exercises that 'combine elements of ancient Stoicism and modern cognitive behavioural therapy'.[34] A recent ebook, *Stoicism Today*, presents selected writings around the project. One section of this addresses questions concerning similarities and distinctions between Stoic attention and Buddhist (or post-Buddhist) mindfulness.

The Mindfulness Movement

While the historical background to the current popularity of mindfulness practices goes back to ancient Indian and predominantly Buddhist traditions of attentional practice, today the practices are largely secular, removed from their philosophic and ethical history. The use of mindfulness in secular settings began with the desire of Dr Jon Kabat-Zinn, a physician working at Massachusetts General Hospital back in the 1980s, to see if mindfulness practices, which he had himself experienced from Buddhist teachings, would be of help to those in his care suffering from chronic pain and stress. To make such practices acceptable in a medical and secular setting, Kabat-Zinn had to delink them from any specific or general 'religious' background and so he developed an eight-week course known as Mindfulness Based Stress Reduction (MBSR). Kabat-Zinn describes mindfulness as 'paying attention in a particular way: on purpose, in the present moment, and non-judgmentally'.[35] It came to public notice with his 1990 book *Full Catastrophe Living* and a Bill Moyers documentary on PBS, *Healing from Within*, in 1993. Later Zindel Segal, John Teasdale and Mark Williams, who were developing a cognitive behavioural therapy approach to the prevention of relapse in depressive illness, trained with Kabat-Zinn and over some years developed Mindfulness Based Cognitive Therapy (MBCT), a form of psychotherapy that brings together Western concern with the mind's contents and Eastern focus on its process. MBCT has been shown in clinical trials to be effective for preventing relapse in recurring depression. Such was the benefit, both experiential and evidential, from these practices that their use became widespread in many different arenas, schools, businesses and other organizations.

The rest of the story is now well known. Success breeds success. Today mindfulness manuals and even mindfulness colouring books lead best-seller lists, and smartphone apps promise one-minute mindfulness. There is hardly a self-respecting large company that has not offered its employees mindfulness training. Indeed, such training is now available for almost any section of the population. Mindfulness is taught for bankers, students, the depressed, the military, not only after engagement for the relief of post-traumatic stress disorder, but prior to engagement to enable them to be more focused soldiers. It has come to the point where reference is made to Mindfulness Lite or McMindfulness, fast mindfood for everyone. This reflects some serious concerns from the side of the Buddhist community about a watering down and weakening of the Dharma: a worry that mindfulness arises from a tradition wherein practice is inseparable from philosophy and ethical stance, and that any mindfulness divorced from these is not true mindfulness. Much contemporary mindfulness training is presented as 'Me Time', in sharp contradiction to its original intention as an opening to 'Not Me'.

Thus there is a concern that attention must be paid to the intention of the practice as well as the actual practice of attention itself, and that merely being concerned with what occurs within the brain is not a sufficient criterion for evaluation. Evan Thompson points out that while mindfulness depends on the brain, it is not

> inside the brain. Certain neural networks may be necessary for mindfulness, but mindfulness itself consists in a whole host of integrated mind-body skills in ethically directed action in the world. It's not a neural network but how you live your life in the world.[36]

On the scientific side there is anxiety that the mindfulness bandwagon is rolling onwards supported by too little and too shallow scientific research, and that popular press reports are ungrounded, which can only result in a subsequent backlash of opinion that could kill useful research and practice as well as the superficial. There is also concern that mindfulness might be a form of 'stealth' Buddhism. Yet there is no doubt at all that mindfulness *works*, that it does help many people – people who would be deterred from the practices if they thought they bore traces of 'religion'.

The concern that the benefits of mindfulness are being so overhyped in the popular media, that the very real claims of the practices are in danger of being lost in over-positive and unrealistic claims, is shared by all those seriously concerned with mindfulness. A popular article citing some twenty benefits of meditation, from 'reducing loneliness to increasing grey matter and helping sleep', inspired Catherine Kerr, a neuroscientist and meditation researcher on Brown University's Contemplative Initiative, to post on her Facebook page in 2014 her worries that such unbalanced coverage could only inspire a negative backlash when inevitably such claims come to be seen as overblown. Her comments evoked an enormous response and the resultant research discussion group of Mindfulness and Skilful Action has become an important rallying point for over four hundred prominent academic, scientific and clinical meditation researchers and some Buddhists. In an interview with *Tricycle* magazine, Kerr suggested that the reasons for an over-positive and uncritical embrace of mindfulness, and the resistance of both scientists and laypeople to worries such as her own, and to those reports that have advised caution or suggested that many research findings are false or questionable, arise from our ever-present desire for certainty and the aversion to indeterminacy. 'Yet', she writes, 'somebody who has a clear scientific understanding knows that the evidence base is always mixed.'[37] It is ironic and rather intriguing, in the light of the Buddhist concern with emptiness and impermanence as opposed to certainty, that the very practice of mindfulness undertaken initially to expose the contingency of our experience is falling foul of the human mind's overwhelming desire for certainty.

Kerr suggests that when promoters of mindfulness only focus on its effects on brain mechanism they miss a big part of the story, just as Buddhist critics of mindfulness also miss something impor-tant when they attack secularized mindfulness as a violation of Dharma. What they are missing, she suggests, is the experiential dimension of what it is like for those in pain to take the MBSR course. The process is not so much about taking away your pain, as about learning how to accept your problem in a new way, about learning to tolerate the uncertainty that is our existential problem. MBSR and its variants help people with this: an answer that should be of comfort to those with a Buddhist outlook. Thus she concludes that our

> tendency to parse the world into competing abstractions –
> scientific reductionist on the one hand or Dharma purism
> on the other – may cause us to miss this hard-to-see qualitative
> shift that may be the true source of the power of mindfulness.[38]

In an article published in the same magazine, Andrew Olendzki, a Buddhist scholar and writer, makes a similar point. He is equally grounded in the experiential, though coming from the Dharmic as opposed to the scientific camp. While questioning the intention of mindfulness training in the corporate world to hone a competitive edge or in the military sphere to make a better combatant, Olendzki suggests that in this age when 'attention is a rare and precious commodity, even as it is spread around so promiscuously, training in attention skills is an understandably popular program'. However, he says that the crux of mindfulness training is to be found in the repeated chorus from the *Satipatthana Sutta*, the locus of meditation training where it states: 'One abides independent, not clinging to anything in the world.' Such disengagement, he says, is incompatible with corporate or military and many other applications of mindfulness in the contemporary world. 'Attending with an unattached attitude that allows us to understand the impermanent, interdependent, and selfless nature of it all is what cuts the attachment and is truly transformative.'[39] It is, at the very least, a disengagement from pain and stress and an ability to live with uncertainty that Kerr mentioned. It seems that common ground may be found if we stick to the experiential. A similar point is made in another thoughtful paper by Mindfulness Based Psychology teacher and MBCT therapist Jenny Wilks: 'The key question is what is the impact on participants.'[40] The gap between the scientific and the Buddhist concerns might be overcome, as indeed that between the popular and the scientific, as long as we remain grounded in the embodied experience rather than spin out into theoretical claims. Indeed the interplay of attention and intention is continuous, arising constantly in all aspects of daily experience.

We must not, however, overlook a further concern with the explosion in the popularity of mindfulness, a concern for how it is taught and who teaches the teachers. While John Kabat-Zinn was fully aware of the benefits and the difficulties of traditional practices, and those like Jenny Wilks are steeped in traditional as well as contemporary knowledge, this is not always the case. Some trainings are being led by those without any knowledge of the traditions of mindfulness, and more dangerously, sometimes without understanding that these are strong practices and presuppose a level of mental and physical well-being. Meditation is, without doubt, contraindicated for many. Someone who has experienced the results and the wrecks of this is Dr Willoughby Britton. Dr Britton's lab researches the effects of contemplative practices on cognitive, emotional and physiological aspects of affective disturbances. In what she has termed 'the Dark Knight Project', Britton has attempted to document, analyse and publicize some of the oft-overlooked dangers of meditation. She points out that while there is a lot of positive data on meditation, no one has been asking whether there are any potential difficulties or adverse effects, and whether some practices may be better or worse suited to some people rather than others. She states that 'Ironically the main delivery system for Buddhist meditation in America is *actually medicine and science*, not Buddhism.' As a result, many people think of meditation only from the perspective of reducing stress and enhancing executive skills, such as emotion regulation, attention and so on.

According to Britton, the widespread popular assumption that meditation exists only for stress reduction and labour productivity, 'because that's what Americans value', narrows the scope of scientific research. The only acceptable and fundable research questions become those that promise to deliver the answers we want to hear, such as: 'Does it promote good relationships? Does it reduce cortisol? Does it help me work harder?' Because studies have shown that meditation does satisfy such interests, the results, she says, are vigorously reported to the public.[41] Less reported are the cases of those who have experienced negative outcomes from such practices. One has only to read the religious literature regarding the 'dark night of the soul', such as its original description by St John of the Cross or the sonnets of Gerard Manley Hopkins,

O the mind, mind has mountains; cliffs of fall
Frightful, sheer, no-man-fathomed. Hold them cheap
    May who ne'er hung there[42]

to know that stages of the journey entail dangers. In the religious quest, however, there is an overall aim or intention that transcends any stages, and there are guides to what may be encountered. When encountering a method that is expected to produce nothing but happiness and enhanced productivity, such dangers, not on the brochure, are neither expected nor supported. Dr Britton is bravely addressing this gap, not only in her research but in practical support for those who have fallen by the wayside. Another excellent interview with her, addressing all the questions discussed above, appeared on the *Tricycle* website.[43] Her plea is for collaboration between those who follow the Dharma and scientific research, free from contamination of belief or expectation on either side. She states: 'If Buddhists want to have any say, they better stop criticizing and start collaborating, working with instead of just against. Otherwise, they might get left in the dust of the "McMindfulness" movement.'

Jenny Wilks is one who without doubt has been collaborating with the mindfulness movement. A long time cognitive behavioural psychotherapist, she also teaches mindfulness and is herself a Buddhist practitioner. We started our conversation attempting to define distinctions between mindfulness, awareness and attention. Jenny suggested that she would 'use the words "awareness" and "mindfulness" almost synonymously but that maybe, attention has more of a sense of being in-the-moment – one-focused – paying attention, and mindfulness, which is also part of awareness, has more of a sense of continuity that you have to remember to have, since it is from the verb to remember. If I look over there and I see a word and I pay attention to it, that doesn't quite seem to be like mindfulness, which is more when I choose to have a flow of awareness of what's round me. Attention can be more like when something grabs your attention, you couldn't imagine that something grabs your mindfulness. I think the differences between mindfulness and attention are quite subtle, but there is something to do with mindfulness being more continuous. When we are miles away and we come back to paying attention again, it is mindfulness that is the capacity to notice that we are not being attentive and as soon as we have noticed it, we are being attentive again. That's maybe more the *sati*, the remembering factor.'[44] We agreed that mindfulness is the practice of attention in a particular, non-judgemental way and could be seen as the practice of trained (or training) attention.

On the ethical question, Jenny felt that the concern that ethics is missing from mindfulness teaching is becoming a 'bit of a view'. A view that she doesn't altogether share. 'Compassion comes out of careful attention. We can train it separately but it's implicit, indeed it's explicit in the concept of *sati*, which according to the *Theravadin Abhidamma* is always an ethically positive mind state.[45] If you pay attention to suffering, you see the cause of suffering in the people around you.' She feels that 'we see ethics too much in terms of Buddhist precepts and my theory is that this is because of our Judaeo-Christian Western conditioning a set of rules, commandments or utilitarianism, and think that perhaps the Buddha's ethics are more like Aristotelian virtue; the question being "what kind of a person are you trying to become?" I think this is what we are trying to do with mindfulness; we are trying to make ourselves into a different kind of person, one that doesn't get depressed, one that isn't stressed, isn't irritable, that is more compassionate. It's not just about rules. So therefore, I don't think that the ethical question is a major thing, and maybe awareness and attention, and a modelling of kindness and acceptance is enough. If you are feeling irritable and upset you are probably not being attentive. It is easy when you are tired to do and say things that are unskilful that if you were attentive you could have prevented, so mindfulness has that capacity, whatever our values are.'

Jenny noted that in MBCT, which is distinct from most mindfulness training and practice, attention is paid primarily to thoughts because it is these that tend to fuel emotion and depression and so on. From weeks four and five of the eight-week sessions, attention is turned to thoughts. Founder John Teasdale hypothesized that the reason for the effectiveness of cognitive therapy is not necessarily because people write down their thoughts alongside positive reframings of them leading to changed action, but more that the very act of writing down thoughts, and keeping a record of them, produces some form of detachment. From this comes an easing of identification with them, which is the therapeutic movement. We are not so much changing the thought, as being able to see, 'Oh that's just a thought – and it comes and goes.' Such disidentification opens up a space for response rather than reaction, the very heart I believe of therapy of any sort, a result of the practice of awareness in any form of healing, whether in psycho or physical therapy. Jenny ended our talk with the thought that such attention to thoughts and thinking is 'not something that I have seen much of outside the world of MBCT or Buddhist practice, the idea of paying attention to thoughts. It's almost as if attention itself might be seen as healing.'

SHINZEN YOUNG is a teacher based in Los Angeles who for several decades has been bringing the teaching of attentional skills out of the traditions, and has developed an impressive package of presenting and teaching mindfulness skills within the context of the contemporary world. When I spoke with him via Skype, he told me: 'I would say if there's one earth-shaking thing that people should know but don't it is that attentional skills are eminently trainable, and I would claim either directly or indirectly impact every dimension of human happiness. To me that's a gospel, the good news. A gospel that doesn't come with any particular doctrine or list of rules. And anyone from any background or culture can benefit by systematically elevating their attentional skills.'[46]

He describes his approach as 'science-friendly and culturally neutral, yet potentially industrial strength in its spiritual impact'. At the foundation of his teaching is his belief that 'mindfulness is not just one skill, it's three skills: concentration, sensory clarity and equanimity.' Within this framework, he presents a clear, logical and embodied teaching that is free from religious doctrine, science friendly and also psychologically sophisticated.

Such is his belief in the power and importance of attentional skill-based psycho-spiritual growth aligned with science that he suggests that enhanced attention might be the next step in evolution. He compared it to me in terms of the ending of the Greek and Roman world views with the rise of Christianity. To this end he has, for many years, been working on the notion of teaching mindfulness skills 'in an interactive algorithmical way, so that means that you have a personal coach who stays with you as you do your mindfulness practice and gives feedback and based on that might make suggestions and modifications and it follows a flowchart where you are looking for the natural windows and walls.'

Believing that this is optimal for teaching and supporting people, it has further occurred to him that 'that algorithm could be automated with multimedia software, thinking of it as an adjunct rather than a replacement for a live teacher . . . a computer program that everyone can have as a personal mindfulness coach available to them 24/7 for little cost, like a vastly improved guided meditation tape or book.'

He has completed the first module, which has been trialled successfully at Carnegie Mellon University. Further trials, using this as an adjunct to psychotherapy, are to be held at Harvard Medical School.

ON THE OTHER SIDE of this explosion of interest in mindfulness, and possibly one source of it, there is also the contemporary worry about the effect that the upsurge of new technology, of new media, is having on our brains, especially on the brains of developing children. Many disciplines must and do come together to consider these topics: philosophy of mind happily today goes hand in hand with science, and under the wide-ranging heading of the neurosciences, we can find cognitive, affective and even contemplative neuroscience. A helpful way forward must be a wide-ranging, non-judgemental and non-instrumental concern for all aspects of attention. There would seem to be no doubt that there is a crying need in today's world for attention to attention. The very popularity of mindfulness in so many forms is evidence for this. There is also a quiet proliferation of the popularity of books with Silence, Stillness and even Emptiness in their titles. Yet we must not ignore the dangers, or run uncritically into the arms of new trends.

Travel writer and essayist Pico Iyer has produced a beautiful little book based on a TED talk called *The Art of Stillness*. Despite his calling as a professional traveller, it is a plea for staying still, for inner rather than outer journeys in our ever-busy worlds. Iyer says that today anyone reading his short book will take in more information in one day than Shakespeare took in over a lifetime.[47] He writes much about 'nowhere', ending the book with 'I think the place to visit may be Nowhere.' But he does not point out something that I realized with amazement quite by chance one day – that

NOWHERE

Is just a space away from

NOW HERE

3

The Neuroscience
of Attention

How we pay attention to the present moment largely determines the
character of our experience and, therefore, the quality of our lives.
Mystics and contemplatives have made this claim for ages, but
a growing body of scientific research now bears it out.

SAM HARRIS[1]

If the Buddhist view gives us the earliest picture of mind, neuroscience gives us the most contemporary one, and one that is
almost daily expanding. What is unexpected perhaps is the extent
of the resonance between the two, though the language is of course
different. For the purposes of this book, two themes are perhaps
the most relevant. First is the belief that the self is not a thing,
a finished permanent, intrinsic entity, but a dynamic process.
The not-self of Buddhist thought is reflected in the processual
multi-centred self of science. The second is the importance of
what Buddhists call practice or cultivation and how this relates to
what neuroscience calls plasticity, the ability of the brain to adapt
and alter in response to habits, patterns of behaviour that instil
consistent patterns of neuronal firings. In the past several decades,
the 1990s being known as the decade of the brain, knowledge
about the way our brains work has proliferated in a previously
unimaginable way. This is not to say that the fundamental questions have been answered. What has come to be called 'The Hard
Problem', that of the seemingly incommensurable relationship
between neuronal brain processing and the subjective 'feel' of
experience, is still far from understood. Indeed there are those,
wonderfully called 'mysterians', who believe it may never be fully
comprehended, being the equivalent of an eye trying to see itself.

Others are convinced that the answer is just around the corner and is entirely material. As noted earlier, yet others believe that the experiential aspect necessary to complete any successful theoretical answer to this hard problem may have to come in artistic guise.

While being obviously central to the hard problem of conscious experience, attention also most fascinatingly bridges the mind/body divide in another form. It is a phenomenon that is understood in psychological terms, yet it has measurable influence in changing phenomena understood in physiological terms. Evan Thompson in his wonderful *Waking, Dreaming, Being* draws attention to one such change. Paying attention (a psychological process) to a stimulus increases the firing rate (a neurophysiological process) of visual neurons sensitive to that stimulus.[2]

One of the most important findings of these years of research is the process of neuroplasticity: the ability of the brain to change its structure and functioning in response to activity and experience. It is now acknowledged that brain physiology and the patterns of neuronal firing continue to develop throughout life, long past the major developmental phases. Moreover, such changes can be brought about by changes in experience both mental and physical. Experience and thoughts *can* become physically instantiated in our brains. It is a two-way process. Just as our genetic physiology affects our experience, so our experience and our thoughts in turn affect our brains. To me this shows in an entirely immanent and unmystical manner how non-material processes, what we could call mind, or even perhaps soul – that is experience, physical and mental – can leave material and physical traces. Perhaps the best-known exemplars of this are research projects that have shown that the parts of the brain used for left-hand movements are highly developed in the brains of violinists, and that the parts of the hippocampus concerned with memory become overly developed in the brains of London taxi drivers.[3] Additionally, it is not only actions that leave traces: it has been discovered that in both action and imagination many of the same parts of the brain are activated.[4]

Most importantly for my subject here, processes of attention have been found to be central to such change, being influentially involved in the synchronization of neuronal firings. Neuropsychologist, writer and teacher Rick Hanson writes:

> Your experiences matter. Not just for how they feel in the moment but for the lasting traces they leave in your brain. Your experiences of happiness, worry, love, and anxiety can make real changes in your neural networks. The structure-building processes of the nervous system are turbocharged by conscious experience, and especially by what's in the foreground of your awareness. Your attention is like a combination spotlight and vacuum cleaner: It highlights what it lands on and then sucks it into your brain – for better or worse.
>
> There's a traditional saying that the mind takes its shape from what it rests upon. Based on what we've learned about experience-dependent neuroplasticity, a modern version would be to say that *the brain* takes *its* shape from what the mind rests upon.[5]

Attention and the Self

If a lesson of neuroplasticity is that we are creatures of habit, it surely behoves us to pay close attention to our habits, to how we pay attention and to what. Neuroscientists and philosophers are united in stating that attention is at the heart of selfhood. Early on in the history of psychology, William James stated: 'My experience is what I agree to attend to. Only those items which I *notice* shape my mind.'[6] Mihályi Csíkszentmihályi, famous for the discovery and definition of the optimal state he termed Flow, wrote that 'to change personality means to learn new patterns of attention', and German philosopher Thomas Metzinger, 'Attentional agency is one of the essential core properties underlying the conscious experience of selfhood.'[7] We might recall the Buddhist psychophysical model of human being as being composed of five aggregates: form, feeling, perception, dispositions or habits, and consciousness. The dispositions or habits are central to the creation of our selves. Self in this model, and here the neuroscientists would agree, is a process – a process of selfing or self-creation. This quotation from Paul Broks approaches this: 'Minds emerge from process and interaction, not substance. In a sense, we inhabit the spaces between things. We subsist in emptiness. A beautiful, liberating thought and nothing to be afraid of.'[8] Here is another description:

> This arising and subsiding, emerging and decay, is just that emptiness of self in the aggregates of experience. In other words, the very fact that the aggregates are full of experience is the same as the fact that they are empty of self. If there were a solid, really existing self hidden in, or behind the aggregates, its unchangeableness would prevent any experience from occurring; its static nature would make the constantly arising and subsiding of experience come to a screeching halt.[9]

Both these quotations come from scientists, the latter one from those also interested in Buddhism, the former from one to whom such teachings are unknown. In another quotation, Paul Broks says:

> From a neuroscience perspective we are all divided and discontinuous. The mental processes underlying our sense of self – feelings thoughts, memories – are scattered through different zones of the brain. There is no special point of convergence. No cockpit of the soul. No soul pilot. They come together as a work of fiction. A human being is a story-telling machine. The self is a story.[10]

We may return to the similarities Evan Thompson delineated between the Buddhist and the neuroscientific descriptions of self-ing noted in the previous chapter. Thinking about oneself as a self and grasping an outside view of self are inseparable. Results of research from developmental psychology, considered in more detail in the following chapter, indicate that these two abilities arise together and build on the capacity for joint or shared attention. This emerges at around nine months of age and arises out of the threefold structure of infant, adult and something to which they both give attention, for example gaze-following, playing and inter-acting with toys, and imitative behaviour. At this time the child gains the ability to monitor an adult's attention to objects, and sometimes the object is the child itself so she begins to mirror the adult's attention to herself, and is thus able to see herself as it were from the outside. As Evan Thompson describes it: 'The self-specifying processes are joint attentional activities that specify each attentional agent as the focus of their shared attention: "I attend to you, attending to me."'[11] Participation in such dyadic activity is crucial to a child's ability to mentally grasp an outside view of themselves and thus to be able to think of themselves as a self.

In time this self develops the ability to shift perspective from the immediate present to past and future scenarios. This process, called self-projection, entails the self being mentally projected into an alternative situation. The sense of self develops through the pro-cesses of memory of the past and prospection (self projection into the future). Thus the autobiographical or narrative self is formed.

One of the more interesting discoveries of the mind sciences is that the brain naturally moves in two distinct modes: focused atten-tion when task-orientated; and what has come to be called default mode, the patterns of firing occurring when not paying attention to different tasks. Different parts of the brain are involved in these two processes. For focused attention activity occurs mostly at the front of the midline cortex; when focused demands on our attention are low and our minds are wandering, default region activity occurs towards the back of the midline cortex. There is also some interesting research that suggests that when in default mode, we are also more concerned with selfing. The brain areas concerned with what we just saw as self-projection overlap with default mode. Evan Thompson says:

> The connection between self-projection and the default network is that during resting or passive situations, when the default network is active, spontaneous thoughts are at their peak, and these often take the form of musing about past happenings, making plans for the future.[12]

Rick Hanson has described this area as 'The Simulator', calling it an 'evolutionarily wonderfully adaptive development in which we can mull over the past and future and try on different possibilities'.[13] Other brain regions, lying along the midline of the prefrontal cortex, have also been associated with both self-projection and default networks. These are seen to be active in what psychologists term 'self-related processing'. This occurs in situations where one is required to evaluate something in relation to how one perceives oneself. While mind wandering can give rise to the creative problem solving and the adaptive functions Hanson mentions, it has also been associated with negative emotions. There is an oft-quoted paper with the evocative title 'A Wandering Mind is an Unhappy Mind' that suggests that when in default mode we are more likely to be judgemental and concerned with failure and thus less happy.[14] It is fascinating to note that this discovery – that the default networks are centrally concerned with self – resonates with the Buddhist concern with the centrality of self. Science sees this as fact, though it has noted that it may lead to unhappiness, to oppose which is the very aim of the Buddhist project, for which thus it is seen as a problem.

Another interesting and oft-quoted paper (Farb et al.'s 'Attending to the Present: Mindfulness Meditation Reveals Distinct Neural

Modes of Self-reference') speaks of different networks that are concerned with representation of the self.[15] In order to explain the feeling of continuity beneath the constantly changing set of experiences, William James posited a 'me' to make sense of the 'I' in the present moment. A neural basis for this distinction has since been found which finds the 'me' in the form of narrative self reference as a function of the medial prefrontal cortices which support self-awareness by linking self references across time maintaining identity. The immediate 'I' in distinction is momentary and present-focused and is based in lateral networks, in particular on the right side of the brain. These networks support open, sustained, present-moment awareness. Studies suggest that these lateral networks are conventionally easily co-opted by the medial narrative mode. Farb et al. suggest that the cortical midline activity which underlies 'narrative-generating mind wandering is very similar to activity associated with the "default mode" of resting attention'.[16] Studying the activity of these networks with fMRI in the brains of those trained with specific training in monitoring moment-to-moment experience (individuals with MBSR training) and a control group, the study found that a default mode of self-awareness may depend upon a habitual coupling between midline prefrontal cortical regions supporting cognitive-affective representations of the self and more lateral viscerosomatic neural images of body state. There is more than a little overlap and also differentiation of terminology around such studies, and it is no doubt incorrect to attempt to make them 'fit' clearly with one another. For example, the philosopher Daniel Dennett refers to the 'autobiographical' self and phenomenologists to 'narrative self', so conversations around the topic tend to be somewhat difficult. However, it would seem that in general terms there are different parts of the brain forefronted in task-orientated and in mind-wandering mode, and that the areas to do with self-projection (past, future and imagination) overlap with those of mind wandering or default. Also that practices of attention, whether focused attention or open awareness, that involve returning the mind from distraction can and do affect default mode.[17]

Conversely there is at least narrative evidence for a loss of self-projection during times of great focus and involvement. Philip Glass writes most interestingly of his own experience of creative attention, speaking of the passage from 'usual ordinary attention to the extraordinary gathering of attention that is required to accomplish something that is unbelievable'.[18] To achieve this, he says, requires a sacrifice. 'What is given up is the last thing left that we are holding on to: the function of attention we use when watching ourselves.' To break through the barrier into something new, we have to give up the ability to watch ourselves.

Mihály Csíkszentmihályi has explored these states of attention in action and would agree with Glass. He says using mental energy creatively relies on the amount of uncommitted attention available. And

> In a person concerned with protecting his or her self, practically all the attention is invested in monitoring threats to the ego . . . Another limitation on the free use of mental energy is an excessive investment of attention in self goals . . . When everything a person sees, thinks, or does must serve self-interest, there is no attention left over to learn about anything else.[19]

He describes how to be creative needs both focused and open awareness, focused to protect against distraction, but open to experience. This is not as paradoxical as it might first appear:

> These contrary ways of using psychic energy share a similarity that is more important than their differences. They require *you* to decide whether at this point it is better to be open or to be focused. They are both expressions of your

ability to control attention, and it is this, not whether you are open or focused that matters.[20]

He then gives some suggestions as to methods of building up the habits that make it possible to control attention: to take charge of our schedule in a way that suits our own rhythms, to take time for reflection and relaxation, to choose and shape the spaces we live in, to pay attention to what truly gives us pleasure and pain and to embrace more of the former and less of the latter.

It might be useful here to revisit his definition of Flow, a state of optimal experience, and indeed attention, often involving difficult activities that entail an element of novelty and discovery. He found that such a state was described by those who experience it in almost identical terms, regardless of the activity giving rise to it. He defines it as being comprised of nine main elements:

1. There are clear goals at every step of the way.
2. There is immediate feedback to one's actions
3. There is a balance between challenge and skill
4. Action and awareness become merged in a one-pointedness of mind
5. Distractions are excluded from consciousness
6. There is no worry of failure
7. Self-consciousness disappears
8. The sense of time becomes distorted
9. The activity becomes autotelic – an end in itself.[21]

I think that Philip Glass's description above of the loss of self-monitoring while writing music and the distortion of time he also mentions would count as prime examples of Flow.

Attention and World

If attention is central to our experience, even our creation of 'self', it is further crucial to our experience of the world. As noted earlier, Iain McGilchrist believes that attention is central to our experience of the world, indeed to the very creation of the world we experience, writing, 'We can only know the world as we have inevitably shaped it by the nature of our attention.'[22] McGilchrist is best known for his hypothesis concerning the two hemispheres of the brain described at length in his impressive book *The Master and His Emissary*. He sees this distinction between the hemispheres in terms of attentional issues:

> Differences between the hemispheres in birds, animals and humans ultimately relate to differences in attention, which have evolved for clear reasons of survival. But since the nature of the attention we bring to bear on the world changes what it is we find there, and since what we find there influences the kind of attention we pay in future, differences of attention are not just technical, mechanical, issues, but have significant human experiential and philosophical consequences. They change the world we inhabit.[23]

He proposes that the two hemispheres are asymmetric for a purpose: survival, how to eat and to stay alive. The first needs close attention to small detail, the latter broad vigilant sustained attention. This asymmetry is important and lies predominantly not in *what* takes place in which hemisphere, but in *how* they work, and that for optimum functioning, the right hemisphere, which sees the bigger picture, should be the Master, with the left hemisphere, whose mode is more focused, utiltarian and abstract, as Emissary.

I had first arranged to talk with Iain McGilchrist following a lecture he gave in East London. However, after a rush hour Tube journey that physically and mentally illustrated the pressures of contemporary urban life that may account for the less than optimal hemispheric balance of which he writes, we finally gave up the attempt to talk in a crowded and noisy restaurant. A couple of weeks later we spoke on Skype from the quiet of our respective homes, he from a Scottish island, I from Devon. At this time he emphasized again the importance of attention, which he sees as not a function, something which one does, but rather an aspect of consciousness. 'In cognitive models of mind there are so many things that are made to sound like something a machine could do – like it could remember things – well of course it can't remember them but it can do a simulation, but it can't even simulate attention. Attention is just the way one's consciousness is disposed at the time, so it is an integral aspect of consciousness.'[24]

When I suggested that attention is at the heart of his exposition of hemispherical asymmetry, he agreed: 'I would go so far as to say that ultimately all the distinctions that I have recorded can be traced back to this difference in attention which is fundamental. The other thing about attention, not only is it an aspect of consciousness but it is an aspect of the world that comes into being through consciousness. So it's completely at the core. If, as I believe, the world is not a lump brute core somewhere out there but is something that to an extent we can't calibrate and we don't know, brought into being by our interaction with it, by our attention to it and by our conceiving it, then attention is not only an aspect of the consciousness we bring but it is also an aspect of the thing that arises from that, which is the phenomenological world. So it's a moral act, attention, as well, because you have a responsibility – how you attend changes what comes into being. One can obviously see it as a moral act in certain circumstances such as caring for people, even being a therapist or being a doctor, just being a husband, wife, parent, the way you attend is important, has consequences for other people, how they perceive themselves, and how they see themselves in relation to you, and so as soon as one attends, and one can't not attend if one is conscious, one's engaged in a moral act. The perfect expression of this which I like to quote is from Louis Lavelle who is a not particularly well-known existentialist philosopher, French – around 1930s – who said "La charité est un pure attention a l' existence d'autrui" – in other words, "Love is a pure attention to the existence of the other." This locates attention very clearly in the moral realm.'

Seeing attention thus as a moral act, I asked if it followed that he would also see it as a skill or ability that can be trained. Again he agreed: 'Absolutely. It stands to reason that would be the case but I just happen to have here somewhere – [searching]. This is from a man called Aric Sigman – he is quite interesting on the impact of screen media and technology on children and attention. He says that it is fairly obvious that "these results", referring to a study of his, "suggest that compassion can be cultivated with training and greater altruistic behaviour may emerge from increasing engagement of the neural systems implicated in understanding the suffering of other people, executive and emotional control and reward processing". In other words, once translated out of cognitive scientese, that aspects of humanity including empathy and ability to see one's way into another person's position can be cultivated, which obviously it can.'

I then asked him what attentional habits or practices he would consider most important for the achievement of hemispheric balance and to counter what he has described in detail as the current over-strengthening of left hemispheric modes of attention. While laughingly disclaiming superior knowledge and stating that 'Here I'm not any better than the next man at saying what one might do because I think my knowledge is about the way we are rather than what to do about it', he did suggest that 'what seems to me to make sense both theoretically and in practice is our dear friend mindfulness, which has become such a cliché that one can hardly mention it. But mindfulness is hugely important. If it's true, as I wrote in *The Master and His Emissary*, and I believe it to be true, new attention, the attention to all new experience that isn't already processed and isn't already packaged, is yielded by the right hemisphere and here one would refer to Elkhonen Goldberg and Costa research on this; that almost every aspect of experience when it is fresh, when it is present or *presencing* rather than being presented, is right-hemisphere dependent, then one would imagine that mindfulness should be a route to enhancing that. And since the book came out there has been research that has shown precisely that: that it does engage widely distributed networks but largely in the right hemisphere. So mindfulness would be one thing.[25]

'At the most elementary level, one of the answers as to what could we do that could help, would be to slow down. That's a very banal thing to say, but it's true that doing things slowly releases an entirely different character to them and to you, which is more rewarding. The whole drift of modern life is trying to pack in as many things as you can as fast as possible. But also time at a much more fundamental level. I've been reading Henri Bergson on time. He is very interesting on the topic, in terms of motion. I think our ideas of extension and time depend on – well not only the ideas perhaps the reality too – depend on motion. If the universe didn't move, it wouldn't have the dimensions it has, or the other way round, but they are intimately connected with one another. The right hemisphere seems to be the one that is able to understand a stretch, or depth or extent of time, as indeed it is the one able to understand a stretch, or depth or extent of space, whereas the left hemisphere seems to focus on a series of points in sequence in space or time, which is quite different and basically, therefore, doesn't have the experience of space or time.

'Perception is very much more dependent on the right hemisphere than the left. In all modalities the right hemisphere is much better at perception than the left. Perception, again because of the cognitive model of a machine registering data, suggests something much inferior to what I believe perception to be, which is a creative act; so that when we perceive there is no such thing as just taking in data, we always take it in as something; we are always creating something out of it. And what we create matters, and the right hemisphere is much better at that. It also has the periphery of the field of attention and it stands to reason that most new experience is likely to come from the periphery and so the right hemisphere always is, as it were, the guardian of the new – what's coming in. But not in the sense of novelty, not in the sense of novelty as a fantasy – novelty as a creation out of things we've got put together in a different order which is what the left hemisphere seems to be engaged in – a wilful creation of new things by taking what we have got and reassembling them in a new way. But to be truly new is to be another thing. In the new book I contrast the fantasy-like novelty with newness; perception which is so austere that you wouldn't notice that something new has happened: just staring at – or staring at is the wrong word as it suggests a different mode of attention – but being in communion with, being attentive to the natural landscape, looking at bleak things like stones and pools and so on – to such an extent that they become newly alive. It's the complete opposite of the squandered wealth of ingenuity of creating novelty, so it's just seeing things for the first time. And the right hemisphere seems to be the one that delivers the sustained open attention to what is. So on all these grounds you would expect the right hemisphere to be the one for which new things arise.'

To my suggestion that many of the helpful practices seem to be around not-doing, receptivity or undoing, he agreed, speaking of a lecture he gives titled 'The Power of No': 'the point being that negation is profoundly creative. We completely misconceive that creation is about going out there and doing something, whereas it's usually about stopping and being silent and creating a space in which something can grow. Because it's not that we can make it happen. We don't have the power to make things happen, we have the power to stop things happening or not stop them happening, and most of the time we stop them happening by our doing. When we stop doing, we then stop stopping things happening, which means they can happen at last.'

Speaking of the recent popularity of books and ideas around silence, emptiness, stillness and space, McGilchrist noted that while such emphasis may be welcome, it may also hide a danger: 'It could form a parallel to what happened in postmodernism: the very welcome idea that every notion of certainty needs problematizing can lead to the too easy step from the fact that no truth is certain to the notion that there is no truth. I think it's just the opposite: if it were certain, then we would know it couldn't be true. It's only in uncertainty that we find truth. I think that rather important positive turn on it can be lost – as it can in emptiness – turned into a kind of free-for-all as we inspect the void and cheer one another up on the way to endless meaninglessness. Of course, you and I are not using emptiness or negation in that way, in some very subtle ways, we are talking about an emptiness that is a fullness, it is not just any old emptiness. Any old space is not good enough. Empty space is one thing, but the empty notes between a great piece of music is quite another.'

Such discussion led on to a discussion of 'soul', which McGilchrist had spoken about in a lecture at the Royal Society of Arts as 'an attitude, a disposition, a certain kind of attention'.[26] Today he adds, 'In other words, it is a way or a process and not a thing.' He explains that it is a process that is out of central vision, which is again the field of the right hemisphere. He had also said earlier that it can only be talked about metaphorically, 'yet is not out of touch with embodiment', which I had found very interesting. He explains this: 'Absolutely, yes I wanted to get away from the idea that the soul was some sort of wraith that had no corporeal characteristics. I think that whatever it is that we mean by it is in the body too and I don't think that is a very foreign idea in many traditions of mysticism; though it is foreign to some Christian traditions that have tended to divorce body and soul, rather following Plato. We must pay Plato his due for being a genius, but on the other hand he is responsible for so many unfortunate things about the Western tradition. A very obvious point is that in the

West we've come to see indirect expression as inferior to direct expression, whereas in fact there are many things that can only be expressed indirectly or implicitly, through metaphors, through narratives and through ideas and images that are pregnant with meaning, and if you unpack it, like explaining a joke, it disappears. So we are bound to use indirect means of expression.'

We ended our conversation by returning to the hemispheric theory, about which McGilchrist said: 'There is a quite cynical and hard to get round and I think pretty robust argument in favour of the hemisphere hypothesis, which is simply this: I don't think there is any neurologist who would dispute that the right and left hemispheres attend differently. The icing on the cake is that you can find it in birds and you can find it in animals. It is not controversial that they attend differently. That much is accepted. The next step is one that a neurologist wouldn't make but a philosopher would, that is equally uncontroversial: how we attend to the world changes the world. And I don't think that anyone would seriously dispute that. Now if you simply put those two things together, the left and right hemispheres attend differently, different attention creates a different world, it follows as night follows day, that they must bring into being two different worlds, two different kinds of world, and if so you would expect them to have various characteristics which follow from the two kinds of attention. Let us see if that is the case. And the answer is, that it absolutely is the case. So that seems so robust that I can't see how you can dispute it to be honest.'

Training Attention

If attention is so central, why is it that we don't pay more attention to it? Neuroscientist Jean-Philippe Lachaux, the author of *Le Cerveau attentif* (The Attentive Brain) asked this very question, and answered it in a similar manner to Iain McGilchrist.

Why should we interest ourselves with attention? Because it determines our perception of the world, our link with all that surrounds us and with ourselves. It illuminates the world and our thoughts, our sensations and our feeling like a spotlight.[27]

He then goes on to describe how for twenty years he was unable to find a popular-level book, which would teach him how to pay attention, until he discovered a text by a Japanese Zen master that pointed him to the Buddhist tradition. He writes: 'I understood in that instant that understanding of attention cannot be only an intellectual matter, but that it must also be accompanied by practice.'[28] He also concedes that there still exists no consensus among the scientific community as to a definition of attention. This is despite William James's statement in 1890 that 'everyone knows what attention is'. As we have already seen, the processes of attention may be delineated on several different levels, neuronal, cognitive and behavioural. However, in general, I will follow William James, as his experiential definition clearly describes the idea of limited capacity, competing foci and the common understanding.

It is the taking possession of the mind, in clear and vivid form, of one out of what seem several simultaneous possible objects of trains of thought. Focalisation, concentration of consciousness are of its essence. It implies a withdrawal from some things in order to deal effectively with others.[29]

Unless accompanied by attention, the information gathered by our senses fails to register in the mind and will not be stored in memory. 'What you see is determined by what you pay attention to.'[30] Neurons are in competition, and which ones register is determined by the strength of the signal, which is a function of attention; the electrical signal of the attended target will be stronger than that of an unattended one. What guides our attention, what involuntarily draws it or what we consider to be relevant is another matter. Attention and intention are intertwined.

A more scientific theoretical description of attention separates the experience into three stages that occur in different regions of the brain: alerting, orientating and executive control that takes care of monitoring and resolving conflicts of attention. The first two stages, being involuntary, will be of less interest to the idea of attention as a skill than the voluntary process of controlling attention, though we must note the power of habit in control that becomes unconscious. Attention is a limited resource, with many demands on its energies. Lachaux speaks clearly about this at a TED talk, 'Attention, Distraction and the War in Our Brain'.[31] Our brain, he says, is divided: we are not single-minded. Our mind wanders and we lose control of it. We pay attention to what we estimate to be relevant, but since the brain is divided into a collection of different subsystems, what is relevant to one may not be relevant to another. The first movement of attention is alerting according to priority maps. Some items are almost hard-wired into these maps, such as flashing lights, faces and texts, and these pull our attention inevitably towards them. If you fixate a little longer the map will orient and alter in line with your current desires and interests. The map will then take into account our likes and dislikes according to our personal reward system based in memory, which then affects the allocation of our attention. However, we can move our attention away from these automatic movements by engaging an opposing executive system in the frontal lobes. Thus Lachaux speaks of three systems competing for our attention: the habit system organized by our long-term habits, the reward system organized by our likes and dislikes, and the executive system, which is more flexible. This applies to all our sense processes, from sight to thought. He also points out that the reward system is very reactive to novelty and information. An alert, such as a phone ringing, will capture attention. Moreover, if we become habituated to a constantly stimulating environment with much novelty, loss of this will produce boredom and, in reaction to this, novelty-seeking behaviour. All these work against focus.

Only the executive system allows us to pay focused attention. As long as we can hold an intention to be present in our memory, attention will remain constant, but it is hard and costly in energy to hold those two systems for long and continued focus on an intended object. This is made more difficult by the proliferation and competition of our intentions, plans and so on in our executive system. This war in the brain leads to distraction. Some degree of multi-tasking goes on all the time: we listen as we watch, we walk as we eat an apple; the real problem arises when we try to focus on two or more demanding activities at the same time. What then do we require to be focused? To stay focused we need a strong executive system with neurons that are able to stay active for a long time. And we need to avoid conditions such as fatigue. We also need an undivided executive system dealing clearly with one thing at a time.

If we know about these divided forces we can use this knowledge of the three systems in order to enhance focused attention. Lachaux ends his talk by suggesting that neuroscience now has the knowledge to support attentional training, in schools for example, that could enhance well-being. What this emphasizes is that to train attention we have to pay attention to attention itself, not just its content or object. Indeed, Lachaux, using the Japanese Tea Ceremony as an example, states that almost anything can be used to train attention, but mostly things don't, as attention is seen merely as a means to a different end rather than as an aim or intention itself.[32]

WHILE PROCESSES OF change are linked to processes of attention through all stages of development, in later life voluntary processes of attention become even more pertinent to any change. When early developmental stages are passed, further change becomes willed rather than natural. Michael Merzenich, one of the leaders of research into neuroplasticity, found that in his research with monkeys, lasting changes occurred only when his monkeys paid close attention.[33] He termed the nucleus basalis and the attention system 'the modulatory system of plasticity', the neurochemical system that, when activated, put the brain into a very plastic state. Once the main neuronal connections have been laid down in early development, there is a need for a stable and less plastic system. Then something called BDNF (brain-derived neurotrophic factor) is released. In sufficient quantities,

> this turns off the nucleus basalis and ends that magical period of effortless learning. Henceforth the nucleus can be activated only when something important, surprising or novel occurs, *or if we make the effort to pay close attention.*[34]

The indications are strong that attention can be trained. Recent scientific studies of meditation are certainly encouraging. It has been found that 'focused attention and open awareness forms of meditation which involve stabilizing awareness while developing meta-awareness of ongoing mental activities, affect the brain's default network'.[35] One study carried out by Judson Brewer and colleagues at Yale University compared brain activity in experienced meditators with that of novice meditators as they practised three different types of Early Buddhist meditation – concentration or focused attention, meditation on loving-kindness and choiceless or open awareness.[36] Experienced meditators reported less mind wandering during the meditations. Moreover, there was less activity in the brain regions of the default network, while brain regions known to be associated with monitoring and control of mental processes were more active when default network regions were active. This supports the hypothesis that experienced meditators, in contrast to novice meditators, coactivate different brain regions during activation of the default network. It also supports the suggestion that mind wandering and self-related processing may be more accessible to control in trained meditators. On the basis of this, Brewer and his colleagues speculate that this kind of accessibility of mind wandering to cognitive control may eventually lead to the ability to change default mode brain activity not only during meditation but in passive resting states. This could minimize uncontrolled mind wandering and negative mood states.

In the Farb and Anderson research into networks associated with self-representation, modes habitually coupled together are seen to become uncoupled through attentional training. The study used fMRI to map brain activity during evaluative narrative focus and present-centred and non-judgemental experiential focus using subjects with or without an eight-week training in MBSR. The individuals who had not undergone the training showed little change in brain activity while doing tasks related to the different focuses. Those trained in MBSR demonstrated significant changes in brain activity when shifting from narrative to experiential focus. The paper suggests that the reorganization following mindfulness training (MT)

is consistent with the notion that MT allows for a distinct experiential mode in which thoughts, feelings and bodily sensations are viewed less as being good or bad or integral to the 'self' and treated more as transient mental states that can be simply observed.[37]

It ends with the statement that 'a growing body of evidence suggests approaching self-experience through a more basic present-centred focus may represent a critical aspect of human well-being'.[38] In relation to such findings, Evan Thompson notes that they suggest that 'it's easier to disengage from narrative forms of self-identification when we have the kind of training in present-centered awareness that mindfulness practices provide'.[39]

Thompson also describes how chronic pain control can be achieved by voluntarily controlling the activities of the brain regions involved in pain perception (anterior cingulated cortex) through seeing the activity of that region in real time in an MRI scan. By intentionally altering brain activity, pain perception can be altered, and the severity of the felt pain lessened.[40]

Norman Doidge, whose book *The Brain that Changes Itself* documented the extraordinary history of the discovery of neuroplasticity, has followed this with *The Brain's Way of Healing*. He describes what can only be read as almost miraculous results from enabling neuroplastic resources of the brain to heal instances of such conditions as chronic pain, Parkinson's disease, the results of traumatic brain injuries, autism and many other conditions. In each of these cases the healing interventions paired mental awareness and activity with the use of an energy source, such as light, sound, vibration, electricity or motion, in order to utilize the brain's own homeostatic and neuroplastic resources. In one of the most arresting chapters on the use of sound to heal, the interventions reveal that, while conscious attention to unconscious process, a willed attention involving cortical areas of the brain, was thought to be essential for repair, direct engagement with subcortical processes can produce extraordinary results. Stimulation of the subcortical areas of the brain through sound can improve brain organization from the bottom up. In other stories related by Doidge, we find many instances of conscious, intentional, focused attention to reset automatic processes that have been damaged.

Ancient and Modern, Neuroscience and Dharma

The Mind Life Organization has, under the inspiration of the Dalai Lama, been bringing together scientists and practitioners for more than two decades to discuss the meeting of neuroscience and Buddhist philosophy and especially practice. Research carried out by Richard Davidson and others, especially in Madison, Wisconsin, would seem to show incontrovertible benefits, both physiological and psychological, arising from attentional training.[41] Davidson has produced findings that 'support the view that attention is a trainable skill that can be enhanced through the mental practice of FA (focused attention) meditation'.[42] Other research has shown that such training not only results in improved attention, but gives rise to increases in positive feelings, both happiness of the self and compassion towards others, and enhancement of the immune system.

I had a fascinating conversation about attention with scientist Susan Blackmore, who comes to the subject from both subjective and objective perspectives, as a scientist and a meditator, and as a writer on both aspects.[43] Talking with me at her home in Devon, she told me that 'There is all this psychological work on attention and lots and lots of theories. It's such a complex phenomenon with so many different sides to it, so many different kinds of abilities involved that even after more than a hundred years of psychology investigating attention, there's no really generally accepted theory.'

She explained that, from the science perspective, something 'rather complicated comes up. I did the very first ever experiment on what came to be called "change blindness", which is if you move a picture like you were moving your eyes then you don't notice even quite big changes. And this started a whole change in the way people thought about the way we see. The whole change blindness thing became part of a change from thinking about perception – visual perception as building up a picture in your head – what you see in your stream of consciousness – to a completely different view, that there really is no picture in the head – that is not how the brain works. So it's very easy for us to imagine and if you read psychology books, most of them and even quite serious neuroscience books even now take it for granted that you open your eyes and what your brain is doing is taking in all the information, getting rid of the mistakes and the movements and so on and putting together a really good picture of the world. Well it isn't doing that. What seems more likely to be happening and the way I think about vision now, is that whenever you attend to some aspect of the visual world – after all there is a sort of picture on the back of the retina, an image – our brain and the whole of the nervous system attends only to certain parts, extracts lots of information from those and chucks the rest away. You could say that – I don't even like the term stream of consciousness because I don't think there is such a thing – but a stream of impressions if you like – there are parallel streams going on at once, lots of them. If I am attending to that piano over there, there is a piano and a kind of background assumption that I can see everything else, and then my attention switches over to you with the same assumptions and we end up with this illusion that we can see the whole room and everything in it. Actually what's happening is the arising and falling away of things attended to. Now of course that sense of arising and falling away also has a kind of Buddhist connotation, when you think about thoughts, or even more difficult, think about self as being something that arises and falls or passes away and is impermanent and so on. But going back to the science of it, what I find fascinating is that the more I practise thinking of the world that way, thinking of my own interactions with the world as being like that, the easier it becomes, and the more laughable it seems that I ever thought that there really was a picture in there, that I was looking at everything. That sort of change in the way you experience the world takes some doing from an intellectual perspective, and I think that all the decades of meditation have probably made it easier, but it's really fascinating because it also contributes to pulling apart the sense of being a subject of experience, being somebody over here who's got a picture in their brain of what's over there. Instead it's "this, this" – all these temporary things that pop up and go away. So that's a big change that's come about for me starting from the intellectual end, from theories going on in cognitive science in I guess it must be the last twenty years now, and my original experiment in 1995 or '96, but it takes a while for these ideas to feed through, but that's one thing that comes to mind when I think about attention.'[44]

I suggested that this must be a rare experience to have come to this understanding from the intellectual end and also to have the meditational experience to back it up. Sue agreed, saying cognitive science now holds a whole range of views, from some version of the idea that vision is not a building up of a picture but merely the gist of a picture, through to 'no pictorial qualities at all, because vision is not building up anything, it's action, it's action-based perception so that seeing is a kind of doing, mastering the sensory-motor contingencies. I would say there are some people there who like me have found that their intellectual work has influenced the way they see the world, but there are a lot more scientists as ever who somehow seem to keep their science in one box and the rest of their life in another. And however clever they are at explaining what might be going on, it doesn't actually change the way they see the world. For me I think the reason I have taken it on board with such enthusiasm and literally tried to train myself to see the world in a way that is compatible with my scientific understanding of the moment is partly because that's in my nature, I don't want to separate out science from the rest of life. Also having trained in meditation for a long time means that I am constantly alert to how the world seems to me, and paying attention to things – glancing over there at the books it's "oh books arising" – that sort of thing. So I think that's two related but slightly different reasons why I take this so much to heart.'

When asked if she believes that attention is a skill that can be practised, she replied: 'I wouldn't say that attention is, but I would say that the training of it is, yes. Attention is a natural capacity of the brain, it has to be. Let's take a cat brain – [pointing to the cat that is just leaving the room]. He pays attention. He will pay attention very carefully and steadily when there's a vole or a mouse and then he will just lie around. Just like a human being, attention for a cat, and for most animals, can either be internally generated –"I think I'm hungry now and I'll wander over and see if there's food around" – or it's externally driven by sounds, noises, scurrying of things, whatever. And that is the same in humans. So both of those capacities are perfectly natural and inherent, so I think what is trained – I'm only slightly quibbling with the way you put it – it's not that you are training attention so much as you are training the capacity to direct that attention. Direct it for what purpose? That is then the question that arises. So you can see some traditions that mean paying attention equally to everything without choice – choiceless awareness if you like. In others it comes down to deep concentration, very, very narrow attention to certain things. I'm sure there are others. So those are different ways in which the capacity for attention can be trained. And there's also the capacity to notice things. You train in some traditions to notice when the attention is drawn away by something else and some traditions [of meditation] I think would say, "Never let it be drawn away, always be attentive," and in others it doesn't matter, it's drawn away, just notice the fact, that's how it is. All sorts of complicated things are going on when people train their capacity to direct attention. I think that if you train to apply attention in one way, e.g. open awareness, that skill transfers to some extent, and in my experience to quite a large extent, quite easily to paying attention in a completely different way, e.g. narrow and concentrated. In my case it seemed to me to indicate that it's a general skill that you are training here, not the specific skill of the type of meditation that is being trained.'

We spoke about the default network, and differences between brain in task orientation and default mode. Sue related her own experience earlier in the day. 'I was out there digging and some people say that they find it easy to be mindful when they are gardening. I don't. I really don't. I have to be concentrated. All I was doing was digging and weeding, heavy weeding. And I've thought about this quite a lot. I have to concentrate on "Now this weed, now that. Where's the fork?" It's very hard to be mindful, unlike just walking along. And I don't even try to be mindful. In that state I think it's more like Flow. One of the exciting things that Csíkszentmihályi discovered about Flow was that self seems to disappear. Now that seems to fit very nicely with default mode idea – that when you are task orientated and certain parts of the brain are concentrated on the task, you don't put any cognitive resources, i.e. attention, on to self, but when there is no really demanding task there's no kind of organization like that, and it flips back into this default mode when, supposedly, this network pops up that includes the self, and that's the unhappy mind bit, because this is when the self stuff starts coming up. But I don't think that our understanding of where all this goes on in the brain is really good enough. Well mine isn't, and theories are changing fast now. But my interest in the self at the moment is in out of the body experience, which has always been fascinating to me having had such a dramatic one 45 years ago. And they now talk about four different aspects of the self – the body schema, which is the most important for out of body experiences; the sense of agency; the first person perspective; and the sense of ownership of the body. Now that's all before you get anywhere near a narrative self or self image or social self or any of those other things, but *those* are in that default network, at least in the temporo-parietal junction and its connections with various other areas. What's fascinating to me is that learning that different bits of the brain – they are all close – are dealing with these different aspects of self really helps to make it easier to sort of allow the self to be not one thing, not a permanent existing thing. There is a sense of "I'm over here – oh that's the first person perspective bit," and there's the sense of "It's my foot!" is a different bit, and so also in this way my scientific investigations relate to my Buddhist training.'

More than most people – scientists or meditators – Sue has pursued the *experience* of selflessness to the extreme, which she describes in *Ten Zen Questions*. She tells me about a seven-week practice of extreme mindfulness to the nature of experience, to paying attention to the present moment without any wandering at all, in which she discovered that 'a sense of self began to go away.

If I really paid attention the streets were going past. I would get a flip from "I am walking the street" to the streets going past, and from that it's very easy to have a sense of stuff as just arising and falling away – not to anybody – it's just doing that.'

Following a recent retreat with Leigh Brasington, and his story of Bahiya described earlier, she realizes that all mindfulness is not so extreme, and has come to a new understanding that she finds very helpful. She now separates 'two completely different states. I can practise what I would call "ordinary mindfulness", which is more like being aware of what I am doing at the moment; being in the present moment in a very light kind of way. I'm just sitting here in this chair talking to you. I'm not wandering off somewhere else, I'm paying attention to this. I'm being mindful. And if something happens over there, I'll notice it. And that is different from when you really push that and really push the attention – and this is relevant to attention – the self drops out of it. There's just stuff arising and falling away as I was describing in an intellectual fashion previously. The stuff just comes and falls away, and it seems more as if the table is kind of arising. It's much easier when walking. It's just a kind of arising and being whatever without anybody it's happening to. So now in my own mind, if I'm doing ordinary mindfulness I'm not upset with myself because I am not doing what I now call "Bahiya way". I think of that as something different, which I only practise when it's easy to do, like going for a walk, some simple thing that doesn't require higher-level control. And that's been a huge relief.' She explains that Leigh is keen on being able to live in the relative, while she has always been drawn to the absolute.

When I asked her about whether a sense of continuity can be retained in the selfless arising and falling away of 'stuff', she replied that 'one answer, and this does relate to attention, is to say that continuity abides in the physical world and not in our self. This really does relate to attention, because one of the ideas in the Enactive theories of cognition, in particularly, sensori-motor theories, is that you get the sense of continuity of the world – I'm not talking about the continuity of self now, but the continuity of the world – because you can always look again, you can always see just in time.[45] So, though I am looking over there at the yellow beak of that furry bird over there sitting on the piano [laughing] I have this impression that the whole world is there, even though it is so blurred – if I keep my eyes still it is so blurred that there is not much there at all – and I am paying attention to the yellow beak so there's not much processing of any of this, but I can maintain, or the system can maintain, the illusion of being aware of all of this because all I have to do is look. Some gist has been retained. I think it has a lot to do with eye movements. I think that all the time we have a map of location of the body – a first person perspective – the books are over there, the door is there – that is kept, that's the kind of gist that enables eye movements to be accurate; that doesn't go away because I'm paying attention to the yellow beak. It's a sort of framework that allows me any time to adjust to this – and this – and this. It appears every time I pay attention. That confidence comes from the continuity of the world, so this system doesn't need to have continuous monitoring of the bookshelf or the door or anything else. It just has this kind of gist of where to look, and that's all it needs, the continuity remains out in the world – not in the mind.'

Changing tack on to ideas of outsourcing our abilities, Sue spoke of her interest in the evolution of technology, citing memes and temes.[46] In 1999 Sue wrote *The Meme Machine*, where she talked about memes as 'originally a parasite turned symbiont; that words and language kind of parasitized our brains and caused us to get bigger brains and changed our behaviour'. Now she believes that a similar thing is going on with a third replicator, which is digital technology. 'It is parasitizing our brains and causing us to adapt our brains to deal endlessly with much, much more information than can ever make us happy. Temes [a term she coined following on from memes] are exploiting aspects of our brain to their benefit so that things that our brains are capable of doing, taking in masses of stuff, computing, writing stories, making images, passing them all around, arguing about stuff, all of these things which create more and more temes to compete with one another and get into more brains, all of that is kind of why our brains are being exploited in particular ways.'

She explains that memes were the second replicators and temes are the third. The first were the mitochrondria, 'initially free-living bacteria that then got absorbed within the cell to make eukaryotic cells. What they did was to outsource protection and food supply and concentrate on the job of supplying energy, whereas the other cells outsourced energy production to the mitochondria and kept on doing the other things, so they both benefited. That's why it's called a symbiotic relationship. I have proposed that with the temes all spreading now in this technology we have created we are heading for a role like the mitochondria. This evolving techno-sphere – we are of use to it for several reasons – one is because we spend endless hours doing email etc., and we are the ones who produce the energy, we dig up the coal, oil, rare metals and so on in order to produce everything that produces and keeps the computers running. And if you look at how the energy graph of servers is going up and up as a proportion of general energy use. I am seeing this technological evolution stuff totally out of our control, evolving as all systems do for their own benefit, changing our role. In that way I think of outsourcing – we are giving up our ability to grow our own food, all the things that use our own bodies in order to spend time in front of screens because that is useful to the temes. All of which is way out of the topic of attention. But maybe not because if you think of attention in the way I would suggest most psychologists do, attention as direction of cognitive resources, or energetic resources in the cognitive system, then this is very much about attention, because what the technosphere is doing is causing us to pay more attention to screens than to our own bodies and how to feed them and keep them healthy and happy. Because in evolutionary terms, human happiness is not the point. In evolutionary terms replication of the replicators, reproduction of the replicators, is important, whether that's genes or memes or temes. The needs for human happiness and health are not those of the world as it is at the moment.'

Sue's talk of temes underlines that fact that while neuroplasticity can be an aid, retraining brain maps in cases of injury or loss, such plasticity can also be a threat as it makes us more vulnerable to external and cultural influences. One has only to look at the enormous rise in the diagnosis of attention deficit disorder (ADD, now more commonly referred to as ADHD – attention deficit hyperactivity disorder) to see this. There is a mass of research about ADD but as yet no clear conclusions about either its increased existence or its relief. What alone seems likely is that it will take a concern with the processes of attention to provide a cure. It is also clear that the old slogan from the 1960s of Marshall McLuhan, that *the medium is the message*, is being seen to be literally true. New media are changing the messages and thus brains.

In all senses, culture modifies brains. There are differences in the brains between those who developed in different cultures and spoke different languages during childhood. Cultural influences may perhaps be seen as cultural habits, and thus reinforce certain pathways just as repetitive actions or thoughts do so. Contemporary rapid advances and changes in media engagement with virtual realities are remodelling our brains. Teachers are noticing that the attention spans of children are decreasing. Television and film producers, realizing this, are directing their programmes and products in this knowledge, and so the cycle is reinforced. Research has linked electronic media and TV watching to the rise of attentional deficit traits. Such reorganization need not be harmful, yet there is disturbing evidence that 'much of the harm from television and other electronic media, such as music videos and computer games, comes from their affect on attention'.[47] The speeded up nature of such programmes feeds an increased appetite for high-speed reactions and affect our orienting response. Need this be harmful? I really don't know, but it is surely something of which we need to be aware. Those suffering from ADHD do not seem to experience increased happiness, nor does their behaviour increase the happiness and ease of those around them. Awareness of the problem is the first step; the healing will surely come from wise use and understanding of the very neuroplasticity that is causing the potential problem. Indeed, Doidge's latest book describes some successful and non-chemical processes that have proved helpful.

Evan Thompson, whose work has already been much quoted above, has recently received a five-year Social Sciences and Humanities Research Council of Canada Insight Grant for a project called 'Attending to Mindfulness'. I think that this is so relevant to my subject here that I would like to quote in full the Project Summary:

> After decades of philosophical neglect, attention has emerged as an important topic in the philosophy of mind. At the same time, the cognitive neuroscience of attention both presupposes and gives rise to foundational philosophical issues about the nature of attention and its relationship to cognition, emotion, and consciousness. My research project will contribute to this convergent focus on attention in the philosophy of mind and cognitive neuroscience.
>
> To my knowledge, my research project will be the first philosophical examination of the significance of recent cognitive neuroscience research on attention training through Buddhist mindfulness meditation, as well as the first philosophical study of the relevance of Indo-Tibetan Buddhist philosophical theories of attention for contemporary philosophy of mind.
>
> My working assumptions are (i) that attention actively structures experience; (ii) that the ability to guide and control attention is essential to being a cognitive agent with a subjective perspective on the world; (iii) that voluntary

mental attention is a trainable skill; (iv) that mindfulness meditation methods train attention in precisely describable and measurable ways; and (v) that mindfulness is a mode of skillful embodied cognition that depends directly not just on the brain but also on the rest of the body and the physical, social, and cultural environment.

My research project has five specific objectives: (1) to develop an account of voluntary mental attention as a flexible and trainable skill; (2) to clarify the conceptual and phenomenological relations between attention, consciousness, and agency; (3) to demonstrate that voluntary mental attention should not be reductively identified with neural processes inside the head, but instead should be understood as a mode of skillful activity of the whole embodied and environmentally embedded agent; (4) to criticize the cognitive neuroscience of mindfulness meditation practices for neglecting the embodied and embedded nature of these practices; and (5) to propose an alternative embodied cognition approach to the cognitive science of mindfulness.

My methodological approach (developed extensively in my earlier work) is to use conceptual and phenomenological analyses of subjective experience to assess cognitive science investigations of human cognition, while using cognitive science models and findings to enrich philosophical analyses. A key new advance (which builds on my earlier work in cross-cultural philosophy of mind) will be to enlarge the domain of phenomenological philosophy of mind to include Indo-Tibetan Buddhist philosophy. A further advance will be to show the significance of this cross-cultural approach for the cognitive neuroscience of attention and mindfulness training methods.

This project will serve as an exemplar for cross-cultural philosophy of mind. It will demonstrate the significance of cross-cultural philosophy of mind for foundational issues in cognitive science. It will also be the first philosophical examination of the significance of mindfulness meditation training for understanding the nature of attention and its relation to agency and consciousness.[48]

NeuroDharma and Brain-based Approaches to Psychotherapy

Rick Hanson is already working with the general public at this edge of the scientific and the contemplative. With best-selling books such as *Buddha's Brain* and *Hardwiring Happiness* and an online training programme, 'Foundations of Well-being', Dr Hanson, who trained as a neuropsychologist, is one of a small number of teachers and therapists who work with the ways we can use the plasticity of our brains to enhance well-being. He explains that the brain is hard-wired to be Velcro for bad news and Teflon for good; a relic of the days when to miss bad news such as a wild animal coming out of the woods meant that no good news would ever be relevant again. In the knowledge of this he teaches methods to enhance the good, techniques for taking time and paying attention to such feelings as gratitude and appreciation. By paying close attention to and expanding these feelings, we may rewire our brain patterns so that they may become the used trails in our brains. On real trails, as we walked and talked in the northern hills of the San Francisco Bay Area, it is easy to feel appreciation and gratitude. Perhaps this landscape has influenced Dr Hanson's thoughts.[49]

He began speaking of the first stage of attention, alerting, as the first conscious apprehension of an incoming stimulus to a nervous system that is waiting for something to happen in a phase of uncertain noise awaiting a signal. If the uncertain noise of the waiting brain may be likened to the surface of an ocean, the incoming signal would be a rock thrown onto the surface causing ripples that are meaningful. While the ocean is chaotic, yet it is full of potential. As only a fraction of our neural substrates are used to represent experience, he points to this unused capacity as a fertile potentiality. Influenced by the work of James Austin and his detailed writings on Zen and the brain, Hanson described to me a link he drew between that first stage of attention, the alerting phase, and Buddhist ideas of emptiness, as the alerting phase involves ancient neural networks that are very much linked to a sense of impersonal rather than self-referential processing. He is interested in the possibility of training the ability to abide in this stage, stimulating this mode free from self reference, that he describes as 'a windshield of now'. This would enable one to abide in the place where everything is, as he described it, 'new new new' and 'fresh, fresh, fresh', before the latter stages of orienting or the mobilization of resources to response occur.

Paradoxically, of course, such training would need the engagement of the orienting and executive phases of attention. Although he points out that the experience of awakening, *satori* in Zen, is often achieved through not-knowing and in a context of surprise, quoting a favourite haiku – 'I scooped up the moon/ In my water bucket – and/ Spilled it on the grass'[50] – he is most interested in training attention to include more of our whole awareness and expanding it to include an ever wider field. Attention may be still focused on a single object, but the focus is widened – for example, when playing tennis the attention is on the ball but widened to include all the court and the opposing player; or when playing music, attention to the instrument also includes the whole band and even the audience.

He is absolutely convinced that attention can be trained, but suggests that it is helpful to look at attention from the perspective of three related aspects or distinct qualities and, in the light of these, to take into account individual profiles to enhance individual practice. These three aspects are the placing and sustaining of attention, the ability to withdraw or screen attention from unwanted objects, and the degree of stimulation seeking. Such profiles may be innate or acquired. He talks of this in terms of jackrabbits and turtles. At one end of the spectrum the most turtle-like will be good at placing and sustaining attention, poor at withdrawing it when necessary, and unlikely to the point of dullness and depression to seek stimulation. Whereas at the other end of the spectrum, jackrabbits with ADHD may well have difficulty in placing and sustaining attention, poor filters for withdrawal of attention and high stimulus seeking, though once engaged in highly stimulating action, such as computer games, their sustained attention may be extraordinarily high. Such people he feels would benefit more from training such as walking meditation, or concentrating on positive emotions such as kindness and gratitude, rather than on the more traditional Early Buddhist types of meditation that perhaps are more suited to turtle disposition. He does emphasize that everyone should take concentration seriously, otherwise much practice can end in merely spacing out.

Rick Hanson offers a weekly online newsletter called *Just One Thing*, which is also the title of a book. A recent posting entitled 'Pay Attention' illustrates his suggestions in this area. Explaining that, because of experience-dependent neuroplasticity, whatever is held in attention has a special power to change our brains, he describes how our moment-to-moment mental process sculpts our nervous system like water sculpting a hillside, so that controlling our attention is the foundation of changing our brain, and thus our lives, for the better. He then suggests seven factors of use in the process of any deliberate focusing of attention, from keeping one's head in a dull meeting to contemplative practice. These are his steps:

1. Set the intention to sustain attention
2. Relax, taking several breaths with exhalations twice as long as inhalation, which stimulates the calming parasympathetic nervous system
3. Think of things that help one feel cared for; that you matter to someone or something
4. Think of things that make you feel safe, thus relaxing the scanning vigilance that seeks for potential dangers

THE NEUROSCIENCE OF ATTENTION

5. Encourage positive feelings such as gladness or gratitude

6. Get a sense of the body as a whole. This sense of things in a whole perspective activates networks on the sides of the brain (especially the right for right-handed people) that support sustained mindfulness, and deactivates the midline networks that are in action when we are lost in thought

7. Stay with positive experience for 10–20–30 seconds as such registering helps to instil the experiences into the fabric of brain and self: 'Neurons that fire together, wire together.'[51]

These are the lynchpins of Hanson's teaching to instil new habits: bringing attention to good and positive feelings and thoughts; cultivating the habit of taking in the good; feeling seen, cared for, appreciated, along with a refusal to indulge in negative thoughts, and a careful analysis to see where we have agency, and what we cannot change. Above all, *practice*.

4
Emotional Attention

Attention without feeling . . . is only a report.

MARY OLIVER[1]

A s one might expect, the neurosciences have also had much to say about the early stages of our development, when our brains are at their most plastic and when there is the most to learn. Recent discoveries have allowed revisitation and reinterpretation of many earlier theories through uncovering the neurobiology that underlies the premise. This is particularly so in the case of Attachment Theory. In the 1960s John Bowlby's writings about early attachment were well known and well respected, teaching the importance of early attachment between a caregiver (usually mother) and child for the subsequent emotional development of the child. However, his hypothesis lost support, attacked by feminist opposition to the centrality of the mother's role, scientific opposition and lack of interest. Luckily its influence on, for example, general hospital rules of encouraging babies and small children to maintain close contact with family during hospitalization had become standard.

At the beginning of the twenty-first century, however, others revisited Attachment Theory and were finally able to demonstrate emphatically that such theory was indeed supported by biological and physical evidence. Professor Alan Schore of San Diego has been the leading figure in this field, and Daniel Siegel and others have carried his work into a more public sphere. A very clear description of the importance of early attachment was given in *Why Love Matters* by Sue Gerhardt.[2] This literature emphasizes a central finding: that the maturation of an infant's brain is experience-dependent, and that such experiences are embedded in the attachment relationship. This relationship clearly demonstrates the link between attention and compassion that Buddhist teachings express. Here it is succinctly described by Dan Goleman:

> The act of compassion begins with full attention, just as rapport does. You have to really see the person. If you see the person then naturally empathy arises. If you tune into the other person, you feel with them. If empathy arises, and if that person is in dire need, then empathic concern can come. You want to help them, and then that begins a compassionate act. So I'd say that compassion begins with attention.[3]

It seems that here we have to make a perceptible shift from an emphasis on attention-giving to an equal one on attention received. What occurs in the early development of babies is what we might call a shared attention between the caregiver and the child. Professor Schore proposes that the development of right brain regulatory processes is central to the regulation of bodily and affective states, to processing social-emotional information and also to the control of vital functions concerned with survival and the management of stress. The maturation of these processes is dependent on experience that is embedded in the attachment relationship between the infant and the primary caregiver. He lays out what he calls a 'transactional model' of development which views the development of brain organization as a process shared between genetically coded programmes and environmental influences. Central to the environmental influences is a dyadic resonance or attunement between caregiver and child that is an empathic communication that forms the foundation of the infant's later ability to undertake for herself her own emotional regulation. The baby picks up not only on the gaze of the parent, but on the emotional cues that they share in an affective synchrony in which in the visual and auditory emotional communications embedded within synchronized face-to-face transactions both members of the dyad experience a state transition as they move together from low arousal to a heightened energetic state of high arousal, a shift from quiet alertness into an intensely positive affective state.[4]

Schore speaks of a right-brain-to-right-brain resonance, suggesting that when two right brain systems are mutually entrained in affective synchrony they create a connective resonance, which is now thought to play a fundamental role in brain organization. He speaks of this attachment process as 'joint attention'. The good-enough mother or caregiver (to use a wonderful earlier phrase from Donald Winnicott), who is in sympathy with the baby's physical and psychological needs, thus regulates the child's states of positive and negative affect, and lays down the basis for the development of a healthy future. Schore proposes 'that the attachment relationship directly influences the development of right brain psychosocial-neuro-endocrine-immune communications'.[5] He describes how, over the course of the first two years, the child's capacities for self-regulation develops from an initial position where the primary caregiver externally regulates the child's affective states, to one where the child is capable of internalizing this function.

Such arousal-regulating transactions, which inevitably include both synchronous and asynchronous experiences, when the mother's desires may not reflect those of the baby, enable the infant to learn emotional regulation. The restoration of positive affect following negative experience of asynchrony may teach the child resilience through trust that negativity may be endured and will pass. Synchronous positive affective interactions will not only create a sense of safety but will also foster positively charged joy, curiosity and self-exploration. According to Schore, the attuned caregiver 'interactively regulates the infant's positive and negative states, thereby co-constructing a growth facilitating environment for the experience-dependent maturation of a control system in the infant's right brain'.[6]

Whether this experience is by and large positive or negative will deeply influence the maturation of brain structure and the psychological development of the infant.[7] All babies ideally require a caregiver who is able to validate and make sense of their experience by responding accurately to what is happening in the present moment and reacting appropriately to their needs. Such experience then becomes physically instantiated and embedded. This period in which the right hemisphere of the brain undergoes a growth process specifically concerned with emotion processing extends from later stages in the womb to the first two years of life. Under positive conditions, early mother–baby interactions at this time facilitate the development of later self-regulatory structures. Positive experience in this period results in brains with richer neuronal connectivity; negative experience leads to altered right brain development and subsequent difficulty in emotional (and attentional) regulation. Early relational experience, which allows for the acceptance of both positive and negative feelings, enables such feelings to be fully experienced and tolerated in adult life.

If, however, such relationships are not experienced, the free flow of feelings, vital for mental and physical health, will be terminated. If these shared experiences are largely asynchronous, the effect on the infant will be negative. Long periods of stress, arising from a lack of attunement to the baby's needs, results in neuronal and chemical disturbances leading to both hyper and hypo arousal that creates chaotic biochemical alterations, a 'toxic neurochemistry' in the developing brain.[8] Profound disruption to the attachment bonding process, such as maternal or other care deprivation or trauma, can produce psychobiological and neurochemical dysregulation and imbalance in the developing brain, resulting in abnormal development of neurons, synapses and neurochemical processes, leaving damage to both physiology and functioning, which has been related to later post-traumatic stress disorder, depression, anxiety, addictions, attentional deficit and immune system injury.

Dan Zahavi, a Danish philosopher of mind, in a more recent paper about sharing and attention, posits three distinct levels emerging in development that determine forms of sharing that differ in content and function. In primary intersubjectivity, appearing approximately from the second month, a kind of mutuality arises, and infants engage in face to face interaction and reciprocal exchange. Secondary intersubjectivity comes at around seven to nine months, when the infant is able to break away from face to face reciprocity and engage in sharing with others in reference to things out in the world, a kind of triadic attention that includes child, carer and mutually attended object. Tertiary intersubjectivity arises at around 21 months, with a transition into an assertion of ownership and self-recognition demonstrating self-objectification and self-consciousness.[9]

When care in these early stages is inadequate, it is indeed likely to result in unhappiness and unhealthiness. But all is not entirely lost. Researches into neuroplasticity, as described earlier, have shown that development continues throughout life. Just as shared attention in early development sculpts subsequent brain paths, so a somewhat similar process in psychotherapy may help to repair the early damage. Just as early positive emotional experience can predispose a child to future emotional stability by creating healthy pathways and negative experiences and lack of care can set up subsequent difficulties in emotional regulation, so care and attention in later life can attempt repair. Linking his work in child development with the field of psychotherapy, Schore states, 'The promotion of affect regulation is now seen as a common mechanism in all forms of psychotherapy. The interactive regulation embedded in the therapeutic relationship functions as a "growth facilitating environment".'[10] Schore has even proposed that 'the emotion processing right mind is the neurobiological substrate of Freud's dynamic unconscious'.[11] Again attention is at the heart of this process. The relationship between therapist and client co-creates a region of safety and attention that facilitates healing by reproducing the early situation with the attuned caregiver that was most probably lacking in the patient's early development. Schore has written that 'A psychoneurobiological updating of trauma theory leads to very different therapeutic approaches that are consistent with updated clinical models in which the primary mechanism of the treatment revolves around relational "non-interpretative" interventions rather than verbal interventions.'[12] He and others have suggested that the modality of therapy should reflect the age at which the major trauma or neglect was experienced. If the damage was pre-verbal and rational, then the cure must access these levels of experience, before understanding can be translated into words and reason.

Talking to a special education teacher revealed her experience of this process. She told me that what the difficult and damaged boys in her care responded to was attention: their bad behaviour was a call to her attention. By refusing to react to this negatively, remaining calm but always consistent, truthful and firm under all provocations, she found that if she rewarded good behaviour with more of her positive attention, slowly but definitively the behaviour problems decreased and she could get to the work of teaching classroom subjects. She said that the first quarter of every year would be taken up with behavioural problems, but once that was over, handled always with calm and consistency, she could go on to teach.

Psychotherapeutic Attention

Attention is at the heart of psychotherapy: first to bring awareness to problems, patterns and pain, second to control unconsidered reactions and replace them with considered responses. Thus awareness practices are central to two Buddhist-influenced psychotherapeutic trainings; that of Naropa University in Boulder, Colorado, and Core Process Psychotherapy in the United Kingdom.

Both these trainings have as their foundation awareness practices and commonly speak of such factors as resonance, exchange and joint attention.

I spoke first of these themes with Maura Sills, the creator of Core Process Psychotherapy, and the director of the Karuna Institute in the UK, now in its 25th year, where she and her staff train psychotherapists in CPP, a form of psychotherapy founded on Buddhist contemplative practices.[13] Maura had been my own trainer as a psychotherapist some years ago. CPP arose from Maura's experience of Western psychotherapy training at Esalen in California with Dick Price, who was the successor to Fritz Perls, originator of Gestalt therapy, and her practice with Buddhist teachers from the Burmese Theravada tradition. Speaking with me recently, she said: 'Core Process is based upon attention and intention. It is founded on just holding attention. When you asked me to consider these questions it took me back to a consideration of the difference, if there is one, between awareness and attention. For me, awareness is just part of our natural condition. Sometimes it is called mindfulness, but I prefer awareness, and it's a sensitivity – for me it's a receptive field – and we can either be in touch with it or not be in touch with it. So for me considering the difference between awareness and attention: attention has more of a quality of heeding the awareness field, heeding what it is we know. For me heeding is the intention. My Buddhist teacher used to say that you don't have to attain the field of awareness, you just have to fall back into it – it's sensitivity if you like, and for me that is a whole body process. In Buddhism of course, we talk about the heart, which is much more about the field of awareness, not the brain, but there is a refinement of heart awareness. Yet for me awareness is not centred exclusively in the heart. If I really allow myself to pay attention to any part of my body, there will be a touching into that knowing field that broadens out. So you talked about CPP coming from both Western and Buddhist traditions, I think for me it is to the extent that we can trust the contemplative practice which joins up whatever awareness and attention might be, then where we direct that attentional awareness is the intention. So it's the kind of aspiration, the trajectory, the choice.'

She explained that in the training, 'we talk about and teach three attentional fields. We trust that the contemplative practice is both psychological journey and spiritual journey, and I don't think we can fully hear what there is to hear without clearing through some of the obscurations, the psychological limitations, constrictions of consciousness, constrictions of knowing. The first stage of what we teach is very similar to *vipassana*. It's the very simple practice of learning how to direct your attention at will into aspects of our own conditions including psychological and energetic states of mind, in the same way as you would sitting on a cushion in meditation practice. The *vipassana* aspect of that, for me, relates to then trusting that we have developed that capacity well enough, long enough, to free it from the direction, the prescriptive direction, and that free attention in the psychotherapeutic relationship starts with yourself.

'Once that is established then you direct and open that attentional field to the other, so that there is an invitation to be affected by the other, while still contemplatively working with, responding to and being with the effects of being open to the other at more and more profound levels. This for me invites us into a different kind of intelligence. So there are the two attentional fields: being with my own experience as I sit here, and being with my own experience as I open to the other's effect on me as I open my sensitivity field. And that sensitivity field becomes very much part of the mindfulness practice, which is really developing opening the sense gates. And as we open the sense gates there is the opportunity of direct apprehension to what is being communicated by the other, as well as of how we are being affected, how we react and how we respond to it. So those two aspects of mind that one can start opening up become available to us to the extent that we are able to practise in the moment in relation to the other.

'However, that still is not enough in my opinion. There is a third attention field, though of course the process is not as linear as my description. There has got to be a training meditation in the field of one's own arising experience for the second field of being affected by the other, opening through the sense gates to the possibility of direct knowing of what might be unavailable through cognitive endeavour. Then there's dropping both of those attentional fields and allowing awareness to expand beyond the object within the relational practice, however we may describe what that is. Quite often it's the invitation to be open to the non-active ingredients, closer, I guess, to what you might term emptiness: to be able to be in contact with aspects of realms of experience out of which the two of you and your relationship are arising. It gives everything a much more interconnected context, so that what's occurring becomes a part of a wider field. For some it's called archetypal, for some it's symbolic, for some it's universal, or transgenerational or ancestral – symbolic territories where the personal arising relational experience allows its ripples to sense what they are rippling into. Occasionally there is a moment where an absolute sense of meaning arises, which is not a cognitive experience. It's an inclusiveness of everything that's arising through the larger field of awareness, whatever you might call it. Without all of that – the relationship of the inner, the relationship with the client and with the field, it's going to be a partial experience. Obviously the client is the particular, is the fulcrum, but there is that aspect of greater mind. Within the confusion of the experience which, when it's only connected with subject and object will always have a conditional aspect, there's a sense that the therapist needs to let go of what has accumulated in the knowing field, and surrender to something larger, then that mind can really flow free in the exchange. Then there may be a different quality to the arising experience, which gives more of a sense of meaning that isn't always related to the personal meaning. It hopefully enables the therapist, the listener, to deepen through the conditions that are being experienced personally and in relationship, and get more of a sense of the quality of suffering, the quality of distress at more and more primitive levels that sometimes go beyond the relational experience. So then the response of compassion can come from a non- or less-avoidant place. It usually reduces the activity of the therapeutic relationship and gives the communications a quality of being more of a kind of transmission (with a small t) rather than an articulated intervention. So there's a multi-level possibility in the room: whether or not there is comprehension of what is actually in that field and being communicated. My own sense is that being open to it means that whatever the mysterious process is, there is a coalescence at some kind of level of all the connections – of the history of the client, the history of the therapist, the history of the world, the history of the family – along with the kind of knowing how that has also affected the possibility of a less constricted consciousness. There is a point where the therapist isn't doing anything but trusting the expanded field.

'Then, of course, there is a coming back in. Because we are human, we can't stay open. Sometimes we are not resourced enough in ourselves to stay open; sometimes the client needs more relational contact. It doesn't look the same all the time, but for me as therapist, I need to wait until I hear silence before I can truly start to meet the client. And as you know our clients absolutely don't have to know any of this.'

I asked Maura to describe how therapists in training were encouraged and enabled to access such skills of attention. I remember from my own time in her training that the method of therapy was referred to as 'joint practice', referring to the ability to transfer the quality of attention honed in meditation to the therapeutic encounter with a client, and also that a central aspect of this process was to enable them to become 'comfortable with unknowing'. She replied that there were many different ways they try to teach these skills: 'Everyone is expected to have a daily sitting practice, so that they are now cultivating their own capacity as individuals to meet arising experience, to come into relationship with it in a skilful way and to continue to go beyond or through what they think is real. Most importantly we teach by demonstration and example. Quite often there will be a session with a student and in that session the student really gets a sense of the presence of the therapist in a kind of direct way and it helps the student settle down and begin to pay attention in the way that the therapist pays attention, so that's a somatic experience. Then, some of these things can be communicated verbally: there is the way of reflecting back verbally, communicating that experience. We use language such as "wondering about", "pondering about", "maybe", "I'm wondering if": language with no certainty, using gentle probes touching into experience. Then there's the sharing of the therapist, such as: "What I'm sensing into just now for myself is . . .". That's how you describe, not the *what*, the outcome of the enquiry, but the *how* that is happening, the *how* that is being affected and *how* you are then able to describe it. So it's through contact and through example.

'There are specific meditational exercises in the training such as *vipassana* and also Tibetan *Kum Nye* exercises. These are somatic exercises in patient endurance that is essential working with all levels of body/mind, and they allow the practitioner to work towards more subtle and subliminal levels of experience, including touching into some of the really early tendencies where there is an opportunity to stop the forward movement of formative experience. Learning that through body practice allows the therapist to develop more of a sense of really being able to sit and be present in the experience of being with clients without that driven need to make a difference.

'We also do relational work, working in twos, in threes and in groups of five and six sometimes. We teach people that if you just allow yourself to offer whole-hearted awareness to the other, as I have described in connection with the three attentional fields, quite often there is nothing else needed. There is a basic premise that awareness is curative in itself.

'We also work with helping, supporting and strengthening people's connection with whatever their experience is of the sacred. Obviously all the people who take the training are not Buddhists, so people word that differently. We tend to call it "source", though I actually prefer "sacred", referring to that which connects us to that which in our nature is not conditioned. We do practices of working with the *Brahmaviharas* (the Buddhist sacred realms of friendship, compassion, sympathetic joy and equanimity), establishing opening up and expanding these feelings through attentional practices either around or in relationship to an individual. It sounds very artificial, and in the beginning it can feel artificial, but gradually, as people get a real relationship to the non-conditioned, or less-conditioned field, then it seems to be that it enhances what is possible in terms of the suffering, the conditioned, the difficult that come into the therapeutic relationship. One is enabled to touch the depth of experience much quicker without getting stuck in the peripheries of the experience. So this lived capacity for the practitioner to trust the universal holding field, or the sacred, the interconnected field, seems to benefit in terms of the work we have to do in the relational and conditional field. It accelerates something, deepens something, clears something.'

We discussed how the therapeutic relationship mirrored or replicated the early joint attention of mother and child. Maura described how the work of Bowlby, Stern, Lake and Fairburn was important to the training and agreed that 'there is a maternal field which is very similar to what is constellated in the therapeutic relationship. That capacity to recognize the other's being nature, which is what Mum does – or sadly what Mum doesn't do quite often – that reflective capacity from mother's connection to her own being and the larger field, is essential, enough of it is essential, so that the other can internalize that sense of who they are in connection with something larger, which was Mum to begin with.'

Such relationship has to give both the child, and indeed the therapy client, a positive sense of acceptance and love and also resilience in the face of inevitable suffering and difficulty The current emphasis in many psychotherapy circles on positive experience needs to be matched with attention to endurance and stamina to be able to be with difficulty and sadness. As Maura says, 'it's a dance', and I remember a talk she gave many years ago to which she gave the title 'Licking Honey off the Razor's Edge'.

KAREN KISSEL WEGELA, who was for many years the director of the Contemplative Psychotherapy and Buddhist Psychology pro-gramme at Naropa University, Boulder, Colorado, and is still a trainer on the course, expressed many similar ideas. This training, founded on the Tibetan Buddhist approach of Chogyam Trungpa, is centred on the concept of 'Brilliant Sanity' – the 'notion that human beings are fundamentally good. Their most basic qualities are positive ones: openness, intelligence and warmth.'[14] The way to uncover this is through awareness, as Wegela explains, 'bringing our own – and our clients' – attention and curiosity to experi-ences in the present moment, no matter what we are doing, so we begin to see through our confusion and start to uncover our innate wisdom and compassion'.[15] The journey of awareness begins with the attempt to touch bare experience, to contact emotions free of narrative – of the story lines that usually accompany them. Attention is directed to mental, physical and energetic experiences, freed from any attachment, preference and intention for the exploration; just to be fully present with whatever is arising in body and mind (body/mind). As with the CPP described by Maura Sills, it asks of the therapist (and thus passes on to the client) the ability to stay present and to be comfortable with unknowing. It is strangely far easier to know and prescribe than to be open and to let be. Just as this reveals the positive healing power of bringing attention to the present moment, Wegela has also written of what she calls mindlessness practices. Mindlessness practices are 'ways of desyn-chronizing mind and body, usually in an attempt to escape from discomfort'.[16] She then uses these in the path to well-being by mindfully employing them. She describes how bringing mindfulness and curiosity to experience has the power to change it.

Although Wegela's practice of psychotherapy, her teaching and her writing are steeped in attention, when I asked her about it overtly, she claimed not to know what attention is.[17] So I asked her what she thought it was, and she replied, 'I always think of it as *paying* attention, so it's deliberately putting your mind on some thing or on your perceptions – it's attending to, being with.' We discussed definitions of attention and awareness and I explained that I was probably using the terms quite loosely in a practical performative, rather than precise, manner, and that I thought the contemplative approaches to psychotherapy with which she had been involved all her therapeutic life were all about attention and awareness. To which she added, perceptively, 'and awareness of awareness'. When I explained that I was interested in the distinction between focused and peripheral attention, she explained, 'In my mind the distinction between mindfulness and awareness is that between (the meditative practices) of *shamata* and *vipassana*.' She preferred to use the word 'wide' or panoramic' than 'peripheral', but agreed that she used both in her work with clients. 'When I ground myself I do it the big way, the wide way. Not just "Where is my body?" It includes the client and the space within which we are meeting.' Pushed further to describe the process, she replied, laughing: 'When I am doing anything well I am not terribly aware of myself. I think the thing I do best, when I am most present, is doing therapy, and there's not a lot of watcher. So I always have trouble telling people what I do.'

I suggested that being so experienced in what she does, she is probably in that optimal state of Flow, and asked what she said when she was teaching others who had not reached such a state. 'I rely on a couple of things. One is giving feedback and the other is demonstration. And then I talk about things, though this can never capture the experience. I talk about the basic skills of listening and attending, letting the client know that you are listening; that you have heard them; how to ask a question – simple stuff.'

When I asked whether she believed that attention is a skill that can be taught, or enhanced, she said: 'One thing that comes to mind is learning how to let go. The way I talk about that with people is being willing for the next moment to happen. Unless you already know how to let go, telling someone to let go is useless. If you were my client and you were very caught up with something, I would talk more about coming back to the present moment and letting go of whatever they were caught up with and do it over and over again. I had a client that I worked with who gets into this litany of all these things that are not fair that have happened to him. We were doing some resourcing. And going to parts of the body that are neutral, redirecting attention into parts of the body that are neutral or even pleasant. He suffered chronic physical pain, and at one point I said to him, "You are getting caught up again, go back to tracking, go back to resource." [Resource is a therapeutic tool, a pre-chosen positive image, which can act as a refuge to return to if painful emotions and thoughts become too overwhelming.] He began to protest and I said, "No, you have to be that strict with yourself, you can't keep letting that happen." I was very strict. I sometimes talk to clients about being ruthless, cutting off their thoughts, particularly people who are just tortured by them. "You can't finish the thought. At whatever point you notice it you have to cut it."

'Sometimes I tell them the raincoat story. I had a boyfriend many years ago who gave me many opportunities to feel jealous. I was driving down 19th Street in New York and a woman crossed the street. She was wearing a tan raincoat. She walked across the street wearing a raincoat. It was a grey day. And the raincoat reminded me of the raincoat that was worn by this woman that my boyfriend had shown an interest in. And I watched how my mind just got whirled into it, and I thought, "Now I can't even look at raincoats, this has got to stop!" And at that point I became ruthless about cutting it – which was helpful. I tell that story partly to tell my clients that I get it, I get how hard and how painful it is, and partly to give them a tool. Ruthlessness. I always start with something more gentle, like "Come back to your breath, come back to your sense perceptions", but sometimes it takes more than that. It takes being ruthless; it takes using mindless practices. Just redirect – let it go – do something else. I don't care what, but if it includes your body so much the better. I told someone to watch funny movies, when they were experiencing great physical pain that no one could diagnose. I told them that I felt they needed some breaks. People teach mindfulness for chronic pain as a way of voluntarily redirecting attention, paying attention to the pain as opposed to the story around the pain. I do that also with people.' Karen agreed when I suggested that the distinction between the pain itself and the story around the pain was an example of the distinction between focused and wide attention, and indeed that between focused attention and default mode.

We then spoke about what the Naropa tradition terms exchange, and I wanted to link both with joint attention and the expanded field of awareness of which Maura Sills had spoken. However, Karen felt that exchange was quite different from joint attention. 'Joint attention is two people paying attention to the same thing so there is a sharing. Exchange is not necessarily about attention. I think a lot of times it's unconscious. Exchange happens. It's just something that happens. It's resonance. And it's not joint; joint implies two separate somethings. Exchange is a natural phenomenon of our non-separation – our non-separateness. Transmission in Vajrayana happens through exchange, it's different.'[18]

This reminded me of Maura Sills's description of experience at this level within the therapeutic encounter as 'a transmission, but with a small t'. I persisted that even if it was unconscious, it was still something that she attempts to bring into consciousness and articulation, to which she answered: 'Right, but the phenomenon by itself doesn't have attention necessarily. It's something that you *can* attend to. If you don't attend to it, if you don't recognize it, it can cause all kinds of trouble – because you think it is yours, or that it began with you. It might not have. It doesn't matter as *now* it's yours (as a therapist) to deal with. And the client may not be aware of it. The example that I always use is that if I am sitting with a client and I start to feel fear, without understanding about exchange, I would think that I am sitting with someone who is scary, when it might be the case that I am sitting with someone who is scared. And there is no easy way to tell. Students are always asking me "How do I know that this is exchange?" and I say, "You don't – you have to explore further." You know your own stuff so if it's not in your repertoire at all, then fine. But more often you don't know.'

We agreed that this demonstrated the importance of psychotherapists having done their own work in their training so as to be better able to distinguish counter transference, their own reactions to whatever the client is presenting, rather than exchange which is picking up on the often unconscious and energetic resonance of the client's process. Karen added that through the exchange, 'people feel seen, feel heard, they feel felt, and many of them haven't had that to whatever extent in growing up and that is healing to have that, even though that may never be talked about, never become conscious it is still helpful.'

I suggested that this acknowledgement, this *attention* in the sense of being attended to, the sense of being heard, was at the very heart of psychotherapy and the healing process. It may then be followed by the bringing into consciousness and articulation of that which was unconscious, which then allows for voluntary response rather than involuntary reaction.

IN CHAPTER TWO, I referred to the mindfulness training software developed by Shinzen Young. He is hoping to have therapists prescribe this as an adjunct to the therapy. The hypothesis is that if the client is trained in attentional skills by a professional coach using inexpensive software, this will enable the therapist to proceed more easily and deeply, faster and more economically. He told me that this hypothesis was to be tested at Harvard Medical School.[19]

I hope that the material above clearly demonstrates in theory and practice the importance of attention, first in early child rearing, then in later therapeutic intervention, should that good early experience have been wanting. Obviously the new knowledge about minds, emotions and attention can well be employed in approaches to education that pay attention to emotional as well as cognitive intelligence, indeed to the inseparability and entanglement of the two modes.

5
Attentive Education

The faculty of voluntarily bringing back a wandering attention over
and over again is the very root of judgement, character and will.
No one is *compos sui* if he have it not. An education which should
improve this faculty would be *the* education *par excellence.*

WILLIAM JAMES[1]

Scientific knowledge of the development and the plasticity of
our brains and our minds inevitably has enormous implications for education, as do the technological and social changes in
the world that we are educating children to inhabit. From the
perspective of attention, I listened to three rather different voices.
First, Matthew Crawford, who provides a political philosophical
critique of the ideal inherent within conventional education;
second, Professor Guy Claxton, an eminent educational psychologist and writer who is closer to the classroom; and third, Professor
Katherine Weare, who has been directly involved with the specific
teaching of mindfulness within the school setting.

Skilled Practice

Matthew Crawford, an American writer and research fellow at
the Institute for Advanced Studies in Culture at the University of
Virginia, is a most interesting commentator on the subject of
attention. Crawford believes that many contemporary problems
with attention arise from the loss and lack of attention given to
the cultivation of skilled practices as a way of engaging with the
world. He notes that both William James and Simone Weil suggested that attention is a habit built up through practice, and that

the struggle to pay attention trains the faculty of attention. He feels that current and recent education has detrimentally taken children away from real hands-on experience into a world of representation, and thus that practical, hands-on education should have a far more prominent role in school. Above all, he truly understands the centrality of attention: 'When you talk about attention, you're talking about the faculty by which you encounter the world.'[2] Such encounters may be enhanced or diminished by the quality of our attention.

Crawford expounds these ideas in his books, *Shop Class as Soulcraft* (titled *The Case for Working with Your Hands* in UK) and the more recent *The World Beyond Your Head*. He addresses the subject of attention in a cultural context, taking the path of political philosophy, as in philosophy as a way of life and politics as politics of the personal. He asks what kind of culture supports good habits of attention, finding that it is most certainly not the one we currently inhabit. If, as psychology describes, attention is both stimulus-driven and goal-directed and is a limited resource, it is of immense importance that we are aware of the often hidden attacks on our stimulus-driven attention, and much concerned with supporting the range, flexibility and skills of our goal-driven attention. He gives a worrying picture of the 'billion-dollar scientifically-informed efforts of manipulation' that daily bombard stimulus-driven attention. He exposes the endless advertisements that attack our attention hourly and increasingly, and fill up all spaces and downtime – the wait on the telephone, the walk through the tunnels of the Tube, even the momentary wait at the machine while we pause to take our card, the bill having been paid. He describes the antidote to this as too often a retreat into the private worlds of headphones, tablets and individual devices. These too are manipulated by those wishing to capture our attention. 'We live in an attention economy,' says someone described as 'a user experience designer for a large social media platform for adults', continuing to describe how people in that industry are constantly talking about how to keep people engaged, and that 'everything about platforms . . . is designed to keep you coming back. They tap into our very basic needs – the desire for social bonding, the fascination with information that is relevant to us.'[3]

Crawford is not only concerned with manipulation but sees that both bombardment and retreat demonstrate a loss of public space for sociability, freedom from unwanted address, freedom of attention and free encounter, which engender a loss for imagination. To counter such a barrage he calls for an 'attentional commons', a 'concern for justice in the sharing of our private and yet public resource of attention'.[4] It is a plea for abstinence or restraint from filling every corner of time and space with manipulative tugs on our attention that constitute no less than a use of our minds as a resource. 'Distractibility', says Crawford, 'might be regarded as the mental equivalent of obesity.'[5]

He sees clearly the connection between attention and sense of self, asking how, in such a world, against such onslaughts on our attention, we can maintain a coherent self. He finds conventional models of self highly dubious, stating that when freedom equals the satisfaction of preference and thus maximum choice, the language of freedom has become the language of marketing. Then, a healthy response requires strategies of self-regulation as an escape from self-fragmentation.

His first book, *Shop Class as Soulcraft*, is subtitled 'An Inquiry into the Value of Work', and it is a plea for a different tack for education. Negatively, it is a fundamental critique of the very models and ideals that underpin our approach to education; positively, a plea for a restoration of hands-on learning, and an encounter with things directly as a remedy for the separation of thinking and doing. The targets of his critique are the '*aesthetics* of individuality' and the '*rhetoric* of freedom' that lie at the foundation of current thinking. These are threads that he develops more fully in the later book, *The World Beyond Your Head*, subtitled 'How to Flourish in an Age of Distraction'. He is disturbed by the heritage of Enlightenment thought in the contemporary arena, which he believes has left us in a world of representations and disengagement. His aim is 'to reclaim the real', in support of which he offers an alternative and more adequate picture of 'how we encounter objects and other people. My hope is that this alternative understanding can help us think clearly about our current crisis of attention, and reclaim possibilities of human flourishing.'[6]

He traces the historical foundations that underlie many assumptions that colour our view of the good or desirable life and the intention of attention and education. Many of these he finds flawed and questionable, centrally those of autonomy and individualism. Tracing their origin to the Enlightenment, to Locke, Descartes and Kant and an understandable response to the hegemony of the Church, he feels such individualism has today gone too far. Freedom then supported intellectual independence from outer authority, but it also began the retreat into individual, internal understanding and the bottom line for Cartesian scepticism, that 'I think, therefore I am.' Thus the standard for truth became relocated – no longer found in the outer world, but moved inside our head, becoming eventually a 'function of our mental procedures'.[7] Attention in the contemporary world is redirected from objects and others out in the world to our internal processes of thought and so we find the disengaged subject adrift from tradition, culture and community in a world of representations. In response to this, his philosophic project is to reclaim and reunite us with the real. He believes that being divorced from the necessities of the nature of things that a world of representations obscures has led to the construction of a fragile self that is unable to tolerate frustration and conflict. It is fragile because it is rigid – untested by the contingency and plurality of the real; a contemporary self 'whose freedom and dignity *depend* on its being insulated from contingency'.[8] Though it is far from Crawford's thoughts, it occurs to me that this contemporary picture is a clear reflection of the central Buddhist project; to acknowledge our unawareness and refusal to recognize impermanence and contingency as being the first task or truth, the gateway to liberation from suffering.

He also critiques current ideas about creativity: that it is some kind of mysterious capacity separated from the sort of mastery acquired through long practice and submission to the necessities of the task; a competence achieved through capitulation to the brute alien otherness of objects or other people that structure our attention. Crawford's is above all a plea for the situated self, in some ways a political philosophic version of the Enactive view of cognitive and neuroscience. It calls for a renewed awareness of what chemist and philosopher Michael Polanyi described as the 'tacit dimension', the unacknowledged background of cultural norms and practices that, usually overlooked, support every action. For Polanyi, every action has a tacit dimension, a subsidiary set of embodied and embedded knowledge. Perception is a 'tacit integration of subsidiarily known details into focal awareness of the object perceived, and therefore a skilful performance'.[9] We attend *from* one set of things *to* another. By attending to all aspects of the object we attain to what Polanyi called 'primary indwelling': 'a use of our bodies, perceptual organs and intellectual framework in contrast to the "re-constituting" or "re-performing" indwelling whereby we comprehend objects and events [as] outside ourselves'.[10]

Resituating the self restores genuine agency in contrast to what, in his earlier book, Crawford described as the deskilling inherent in everyday contemporary life. His plea is for genuine individuality and agency, which comes from skilled practice and 'the experience of seeing a direct effect of your actions in the world, and knowing these actions are genuinely your own.[11] Such agency arises out of what he calls ecologies of attention: highly structured patterns of attention engendered by skilled practices, which locate us within a relationship with things and other people 'outside our own heads'.[12] In such an ecology 'the perception of a skilled practitioner is "tuned" to the features of the environment that are pertinent to effective action'.[13] Thus genuine agency is seen to arise 'not in the context of the choices of a sovereign self freely made (as in shopping) but rather, somewhat paradoxically, in the context of *submission* to things in their own intractable way, whether the thing be a musical instrument, a garden or the building of a bridge.'[14] An example of the ecology of attention Crawford gives is that of the collaborative improvisation of musicians.

Thus in opposition to the more general contemporary appreciation of freedom rather than situatedness, Crawford offers an alternative view of human flourishing as 'a powerful independent mind working at full song. Such independence is won through disciplined attention, in the kind of action that joins us to the world.'[15] Discipline is provided by the constraining circumstances of the situation: embodiment, social nature and the historical moment, the ecologies of attention that are always present in skilled practices. Linking subject to world, he sees the embeddedness and relationship of our attention and actions to the 'affordances' the world offers, using the term of psychologist J. J. Gibson and his concept of an 'ecological niche', the aspects of environment that are meaningful to an animal in its way of life. Gibson suggested that what we perceive are not pure objects from a disinterested perspective but rather 'affordances' those factors that the environment 'offers' for our particular purposes.

This understanding of the importance of constraints was also the view of Polanyi, who said that the 'personal is precisely the subjection of oneself to requirements and a reality beyond the self'.[16] There are several paradoxes here. First, that to find a true understanding of self one has to acknowledge constraint and interdependence – a dependence on world and circumstance and contingency. Second, to acknowledge that the world of representations that one has erected in order to secure the self *against* contingency is itself unstable; as Michael Polanyi pointed out, 'all language and symbolic representations have a margin of indeterminacy'.[17] Even Wittgenstein said that the unwritten part of his

Tractatus was the most important part: 'Of what you cannot speak you must remain silent.' Yet the unfindability of rock-solid truth need lead neither to nihilism nor relativism, because the idea of truth is upheld with an Indra's network of embodied, tacit, sensori-motor practices, language (games) and culture.

Thus in the view that Crawford is espousing, like the one described by Evan Thompson, the environment is constitutive of the self, and attention is at the core of this formative process. When we become competent at some particular practice, our perception is disciplined by that practice, and through the exercise of that skill, the self that acts in the world takes on a specific shape. 'It comes to be in a relation of *fit* to a world that it has *grasped*.'[18] In the world of skilled practice, disengaged autonomy is balanced if not replaced by empowerment through submission and acknowledgement of constraint. In *Shop Class as Soulcraft*, Crawford makes an interesting distinction between those arts in which we can be fully effective and those he calls, after Aristotle, *stochastic*. These are the arts in which one is dealing with things not of one's own making, which may never be known and completed in an absolute manner. Such projects – and medicine is an example – will always contain the possibility of failure. They therefore require a distinct disposition towards them, a disposition he describes as both cognitive and moral. In such arts, 'Getting it right demands that you be *attentive* in the way of a conversation rather than *assertive* in the way of a demonstration.' He continues, 'the mechanical arts have a special significance for our time because they cultivate not creativity, but the less glamorous virtue of attentiveness. Things need fixing and tending no less than creating.'[19]

Interestingly one can find echoes in this of the writings of Martin Heidegger about teaching, which he stated 'was more difficult than learning' because

> what teaching calls for is this: to let learn. The real teacher, in fact, lets nothing else be learned than – learning . . . If

the relation between the teacher and the taught is genuine, therefore, there is never a place in it for the authority of the know-it-all or the authoritative sway of the official.[20]

Thinking, Heidegger, wrote, is a craft, a handicraft, 'man's simplest, and for that reason, hardest, handiwork'.[21]

Such views are obviously at odds with a culture that is obsessed with brain rather than mind, with the disengaged self, individualism and free choice and the teachings of self-help manuals. It is, however, entirely in sympathy with the Enactive and the Sensori-Motor views in cognitive and neuroscience, first described by Francisco Varela and his colleagues in 1991, and continued in the work of Evan Thompson, much cited in earlier chapters. Here we find the mind expands beyond the skull, beyond the body, into the world, as Crawford says, beyond the head. It is the view (with distinctions of emphasis) that Andy Clark, Sean Gallagher Alva Noe, Kevin O'Regan and Dan Zahavi have addressed. The writings of George Lakoff and Mark Johnston also share in many aspects of this outlook.[22] It points to the world of engagement and participation that philosophically comes through the work of Heidegger and Merleau Ponty, and it is the *Lebenswelt* of phenomenology.

Following on from the ecology of attention in our relationship with objects, Crawford turns to the 'ethics of attention': the way in which we are alive to the concrete particularity of other people.[23] We are, at birth, as Heidegger wrote, already 'thrown' into the world. We are *Dasein* – being here – already in a world. Chapter Four outlined the concept of joint attention: the dependence of the infant on her caregiver for her recognition both of herself and of the world. We live in a shared world and most often we act together: social interaction is vital to our development from cradle to grave. Yet so much of our implicit philosophy ignores this, and gives us the picture of the disengaged subject. In contrast Crawford and many other of the authorities cited here show that to be truly an individual is only possible within a background of genuine connection.

Finally Crawford outlines the 'erotics of attention' – the attraction of objects that appeal and provide positive energy. While education necessarily requires 'a certain capacity for ascesis', yet, in contrast to the ascetic aspects of struggle to train attention stressed by William James and Simone Weil, Crawford calls attention to the fact that more fundamentally it is erotic. 'Only beautiful things lead us out to join the world beyond our heads.'[24]

> It is not by freely 'constructing meaning' according to my psychic need and projecting generous imaginings onto others that I escape my self-enclosure. It is by acquiring new objects of attention, which is to say, real objects of love that provide a source of energy. As against the need to transform the world into something ideal, the erotic nature of attention suggests we can orient ourselves by a selective affection for the world as it is, and join ourselves to it.[25]

The next words after this, a section heading, are 'Reclaiming the Real'.

Reclaiming the real through the encounter with things and others implies a different aim for education. From education under the sway of the Enlightenment ideal of the disengaged subject and in thrall to a world of representation, Crawford pleads for a more 'hands-on' approach to education, an awareness and direct encounter with things that are more vital than mere representation and abstraction and in contrast to what he sees as the artificial learning environment of schools. 'To reclaim the real in education would be to understand that one is educating a person who is situated in the world and orients to it through a set of human concerns.'[26]

The Creative Classroom

Professor Guy Claxton has much sympathy with the views
expressed by Matthew Crawford. In a long career in psychology
of education he is at the forefront of the introduction of different
approaches to education in the twenty-first century that will facilitate the development of psychological characteristics that are
judged to be of the highest value to young people in today's
demanding world. A strong scientific rationale lies behind the
choice of these characteristics and also the fundamental belief
that they are capable of being systematically developed. Talking
with him before he introduced Matthew Crawford at the Royal
Society of Arts in London, in a noisy café surrounding that challenged our own attention, he explained to me that he believes
attention is absolutely a skill and one that can be cultivated and
which forms an important part of good education.[27] He said that
he tends not to foreground mindfulness or explicit awareness
practices but that the making and reshaping of attentional habits
forms part of a package of tools for the advancement of good
learning habits. He sees himself in the business of teaching in a
way that gradually, over time, forms and shapes certain kinds of
mental, social and emotional habits that are conducive both to
learning and to a good life. Attention is one aspect of that, as is
being a good collaborator, a trustworthy sounding board to
other people, a good evaluator and appraiser of your own product, and being someone who enjoys the process of looking at
what they have done and how to improve it. There is a clear
overlap here with the qualities Matthew Crawford describes as
the mindset of a craftsman. These dispositions, he explains, can
be shaped over time but not taught explicitly in the sense of
sitting people down and doing a workshop or giving a lecture.
'It's much more about how you design schools,' he says, 'so that
they become more effective incubators of these habits. It is not
so much a question of instilling a collection of separate habits

but of enabling a balance and fluidity between different competing habits, such as ambient and focal attention and the ability to switch between them – to enjoy a lively MTV video and also to sit down and become engrossed in a good book for three hours. You can have both: they are not antagonistic but different modes between which one may learn to toggle with a sense of fluidity, and a greater sense of options as to which instruments in the mental orchestra you choose to have play at any one time. And a lot of these things link together, for example, knowing the balance of being good at being sociable and being good at being solitary.'

He gave an example of a university professor of his acquaintance, who says that knowing whether his study door is open or closed is essential to the way he works. If the door is open he is available for conversation; if it is closed he is working silently and is not to be disturbed.

Claxton described how, 'whilst you are helping kids to add fractions or explain the machinations of the Tudor court or whatever else they are required to learn, you are also making very explicit that the deeper part of education is about stretching and strengthening mental and social capacities: so collaboration, empathy, imagination, perseverance, and concentration are presented as learning muscles which the kids can stretch and strengthen and around which the teachers can orchestrate their lessons so the development of these capacities has greater prominence. Usually educational discourse is obsessed with two dimensions – content and assessment – what we are learning and how we know if they've got it. Two dimensions define a plane, thus we can call this Flat Education, but there is always this other dimension in which mind training goes on in the classroom, and you can teach content such as differential calculus in a way that instils passivity and compliance and credulity or you can teach those subjects in a way that teaches critical awareness, collaboration, imagination and independence. It depends on how you do it. And that dimension is always there so, paying attention to it, you get solid education.'

Claxton explained that in the classroom it is often more useful to approach attention from the reverse direction than that usually associated with mindfulness, asking, 'At what point does mindfulness get lost? So we are talking about helping kids learn the ability to manage distraction, in other words to become more mindful of distraction; from the ability to stick with whatever task they are undertaking to developing wake-up moments when they bring themselves back to the task in hand. It is similar to mindfulness, but the focus is not something like the breath but the task in hand, and it becomes a kind of game or challenge to pick up ever more swiftly on distraction and return to the task. We have a number of primary schools where the kids rate themselves each fortnight or so on a scale of 1–10 on how good they've been on resisting distraction. And if they've achieved, say a 7 one week, then they will try to be an 8 next week.'

Claxton described a practice invented by a primary school teacher in Suffolk, now widely used elsewhere, where, when a distraction or potential distraction has occurred, the teacher will say to the class, 'Just stop for a moment and show me your distraction fingers.' The class will have learned this code, and they hold up the number of fingers that correspond to the way they related to the putative distraction. One finger = 'I don't know what you're talking about, I was so absorbed in my work that I didn't notice anything.' Two fingers = 'I was vaguely aware of something, but it didn't really affect me.' Three fingers = a minor distraction, four fingers = a major distraction and five fingers = 'I *was* the distraction!' The object is to get the kids interested in their own distractibility, framing it as something they can learn to gain greater control of. 'It is not,' he went on, 'about techniques for gaining one-pointedness, it is simply getting them to be interested in the process. It is as if there are knobs behind a kind of opaque screen, and this is just about being able to take away the screen and see that I have some opportunity to control my perseverance. It's a complement of the *what* and the *how* of learning. We were always told to "Pay attention" but no attention was paid to *how* to do so.'

Guy gave another example – a piece of writing by a hyper-distractible boy who, if the class was set a task to do a piece of writing, half an hour later on a good day might have written his name. 'His teacher, in a moment of genius, had a conversation with him, and asked him to play a game with her. She said, "I am going to draw a line across the top of this paper, calling one end of the line completely distracted and the other end totally focused. And all I want you to do is every couple of minutes or so to put a little mark on that line to indicate where you are at that moment." Simply, this device enabled him to write ten lines rather than merely his name, because there was a visible reminder as to what the chore was and it became an interesting challenge to see if he could continue moving his line towards the focused end. And finally he drew his own line across the paper without any prompting from his teacher and began making his own marks. It is all concrete and embodied.'

Building Learning Power, the company Claxton has created, presents two frameworks, each of four palettes. The first framework introduces a picture of the ideal powerful learner; the second is a route map for schools and teachers to enable these dispositions. The first framework offers four divisions: Resilience, Resourcefulness, Reflectiveness and Reciprocity, all of which present different aspects of attention and awareness, although Resilience most overtly addresses attention, being subdivided under headings of noticing, managing distractions, perseverance and absorption. Resourcefulness adds to this with reasoning, imagining, capitalizing, questioning and making links. Reflection builds further with revising, distilling, planning and meta-learning, while Reciprocity includes the more social and interpersonal dimensions of imitation, empathy and listening, collaboration and interdependence.

The second framework, the Teachers' Palette, maps the aspects of school and classroom culture that will help to cultivate these habits and build up a whole culture that nurtures the development of the desired dispositions. It has divisions of explaining, commentating, orchestrating and modelling, all of which overtly support the four Rs of the first model. Illustrating this multi-perspectival model, Claxton spoke of talking to students about role models of struggle, people who only achieved after a lot of angst. Demonstrating the importance of the actual design and use of the classroom, he advocates putting quotes on the walls that reinforce the idea that nothing comes easily, and that all their heroes only succeeded because they stuck at it and made many mistakes en route to success. He quoted Michael Jordan as saying: 'The only reason I am as good as I am is because I made as many mistakes as I did'; and Billie Jean King: 'I look on losing not as a failure but as research.' He emphasized the idea that it is important to take away the idea that 'bright' means being able to do things correctly the first time without any struggle. He talked of banishing erasers from the classroom as they symbolize a kind of shamefulness about not having got it right first time, suggesting that mistakes should be highlighted, not erased, and seen as friends that teach the way. Similarly, teachers are encouraged to talk about their own struggles: how it took them time to learn and to admit what they are still not sure about. It takes, he said, a whole raft of things, from what you put on the walls to how you construct activities in the classroom.

The best learning happens in that middle zone between boredom and too easy and anxiety and too hard – the state that Mihály Csíkszentmihályi called Flow. Constructing the activities so kids learn how to monitor their own state of engagement and how to change the activity in order to get themselves back into the zone is key. He suggested that instead of giving them a page of problems to work at in a maths lesson, for example, they work in pairs to write their own problems or set problems for one another. They can then choose the level of difficulty in the problems they are working on. 'If I am working with you and I'm going to set a problem, I ask you how hard you want it to be. Then you say, "Well I'm pretty good at multiplying two number by two number digits together, so I would like you to give me two digit number multiplied by a three digit number and I will see if I can do it." So the students themselves are able to adjust the level of the task. They are co-constructors of their own learning experience.'

It would seem that this authorizes the students to take over some of their own monitoring and that enhances engagement and a self-empowerment that come from the authority being no longer external and imposed. Monitoring is central to attention. To assess how distracted you are builds up awareness, self-evaluation and self-reflectivity, the ability to be self-aware in the moment and to see what needs to happen next. It seems that if you enable one thing, it is has a kick-on effect; by bringing in self-awareness, you bring in ownership so you gain a kind of genuine authority (again the quality Crawford writes about in relation to skilled practice). Claxton sees all this as a learnable process, but suggests that it occurs only gradually, and that the teacher has to see herself as a coach rather than an authority, not doing too much but letting the students lead the process and the review of it.

THE POWER AND VALUE of slow attention in higher education has been explored in a fascinating essay, first given as a presentation at the Harvard Initiative for Teaching and Learning conference. Art historian Jennifer Roberts described how essential she has found it to give students experience in modes of attentive discipline that run directly counter to the high-speed, technologically assisted pedagogies that are current. She wished to create opportunities for students to engage in 'essential deceleration, patience and immersive education', arguing that these are the kind of practices now most in need of being actively engineered by faculty as they are no longer naturally available. 'Every external pressure, social and technological, is pushing students in the other direction, toward immediacy, rapidity, and spontaneity – and against this other kind of opportunity. I want to give them the permission and the structures to slow down.' In her courses, both graduate and undergraduate, students are required to write an intensive research paper on a single work of art of their own choosing. The first requirement for this is to spend a full three hours looking at the work, noting observations and the questions and speculations that arise from such long and close observation. This long attentional exercise is to be undertaken in a museum or gallery setting, thus removed from the student's everyday environment and distractions. Roberts describes how the students are initially resistant, but that after undergoing the experiment, repeatedly tell her that they have been astonished by the potential unlocked by this process.[28] It was this essay that inspired writer Ruth Ozeki to experience and document her own attempt at this process, which I will describe in a later chapter.

Mindfulness in Schools

Katherine Weare is emeritus professor of the Universities of Exeter and Southampton and has been working on developing and evaluating programmes that support mindfulness in schools. She defines mindfulness as 'the ability to direct the attention to experience as it unfolds, moment by moment, with open-minded curiosity and acceptance'.[29] Such practice enables those who have learned it to be better able to be with their present experience and thus respond more appropriately to whatever occurs in the moment. She contrasts this with the normal state of mindlessness in which one moves through experience, rarely noticing the present moment, ruminating on the past or worrying about the future at the mercy of quick judgements coloured by habit and disposition. As she describes it:

> Over time participants who practise regularly report that
> they gradually learn to sustain and focus their attention

for longer periods of time and accept their experiences in a more curious, interested and open minded rather than a judgemental way. They discover how to use felt physical sensations of the breath and the body as 'anchors' to return to when their minds wander and ruminating repetitive thoughts take over. They come to see that thoughts are mental events rather than facts and can be allowed to let come and go, rather than turning into distractions that pre-occupy the attention. This realisation helps loosen the grip of habitual, mindless activity and produces less reactivity and impulsiveness, and a greater ability to examine thoughts more rationally and experience with greater acceptance and kindness. This gradually modifies habitual mental and behavioural patterns which otherwise create and maintain negative mental states, such as rumination, stress, anxiety and depression, and makes for greater mental stability, calm, acceptance, appreciation of what is rather than hankering after what is not, and thus higher levels of happiness and wellbeing.[30]

The Mindfulness in Schools project, created by Chris Cullen and Richard Burnett, supports a programme, .b, which stands for 'Stop, Breathe and Be!', the core of a ten-lesson course for young people aged eleven to eighteen, usually delivered in the classroom.[31] The course is based on the core mindfulness principles of MBSR and MBCT. Each lesson teaches a distinct mindfulness skill. It includes a brief presentation by the teacher aided by visuals, film and sound and practical demonstrations and exercises designed to make the ideas vivid and relevant to the students' lives. The first lesson is called 'Playing Attention' and refers to training the muscles of the mind, likening attention to a puppy that needs to be trained. Lesson two explores different mind states and teaches that 'anchoring' attention in the body, alongside the cultivation of curiosity and kindness, can be calming and nourishing. Subsequent lessons widen the practice and lead from reaction to response in different arenas, mental and physical. This includes a lesson centred on the idea of taking in the good – allowing space for gratitude and positive thoughts. Practices include sitting still watching the breath, awareness of body, mindful walking and awareness of how the body feels under stress. Brief practices are suggested for trying out at home during the week, and the whole course is supported by a student handbook. The aims of the course are given as greater well-being, the ability to fulfil potential and be more creative and relaxed, both personally and academically, and to improve concentration and focus, again both academically and in paying attention and listening to others. It is also intended to enhance the students' ability to work with difficult mental states and stress. A similar course, *Paws b*, has been developed for younger children of primary school age, seven to eleven years.

While there have not to date been sufficient thorough and systematic research projects to evaluate the success of these and other such schemes, research that has taken place into their effectiveness does suggest that they have the potential to fulfil these goals.[32] Mindfulness in Schools has an ongoing research programme that is linked to the Oxford University Centre for Mindfulness. It was one of the examples of programmes singled out by MP Chris Ruane when in December 2013 he addressed the UK Parliament in a speech entitled *Mindfulness in Education*, advocating such efforts to bring mindfulness training into primary and secondary classrooms. There is currently a UK all-party parliamentary group on mindfulness working with the Mindfulness Initiative, in partnership with the four mindfulness training and research centres at the universities of Oxford, Exeter, Bangor and Sussex, exploring the role mindfulness could play in schools, in the National Health Service and within the criminal justice system. In their report, *The Mindful Nation*, published in October 2015, the Mindfulness Initiative made key recommendations: in summary, that in health, access to MBCT should be substantially widened for adults with a history of depression; in the workplace, that public sector employers set up mindfulness pilot projects which can then be evaluated; in education, that mindfulness in schools should be made a priority both for development and research; and that pilot projects to identify appropriate forms of mindfulness teaching should be set up in the criminal justice system. In a newspaper article with the hopeful title 'Why We Will Come to See Mindfulness as Mandatory', Madeleine Bunting, the co-chair of the all-party group and co-founder of the Mindfulness Initiative, argued that mindfulness 'is probably the most important life skill I am learning' and that 'diligently practised, it very quietly and slowly revolutionizes lives in multiple ways – sometimes small, sometimes big. And when you start noticing that process of change – both in yourself and in others – it is quite simply astonishing.'[33] Should these approaches become more widespread, as indeed seems to be the current possibility, classroom calls to 'pay attention' may become far more meaningful, nuanced and life enhancing.

PART TWO *Attending Creatively*

... as an artist. It's about paying attention, that's all it is.
Works of art are just ways to pay attention to different things
and to appreciate what's there.

LAURIE ANDERSON[1]

To this point I have mostly been considering the backstory
of attention, its development, neurological foundations and
the ways in which it might be trained. From here on, while still
concerned with these aspects, I want to turn towards the results
of trained attention. Artists are among those, I believe, who pay
attention most closely, most creatively. In all fields of practice,
artists have trained their attention and their imagination to aid
the rest of us, their audience, readers, listeners and spectators, to
see differently, to follow where they have gone before.

Sometimes attention takes its inspiration from the world around
us; sometimes it is attention to the inner world, to strange forms
of imagination, to the ideas and inspirations that for most of us
fly by on the wind and are lost with the next breath. Such inspirations are attended to, tended, cared for by those we call artists:
attended to with such care that something is revealed or created
in a way that it has never been revealed before.

George Steiner expresses beautifully the dialectic of creation.
'Deep inside every "art-act"', he writes, 'lies the dream of an absolute leap out of nothing.'[2] This dialectic between emptiness and
form plays out in all the arts: as music, words and dance emerge

out of silence and space. Steiner is also deeply conscious of the crossing point between artist and audience, seeing it fundamentally as no less than an existential project:

> Responding to the poem, to the piece of music, to the painting, we re-enact, within the limits of our own lesser creativity, the two defining moments of our existential presence in the world: that of the coming into being where nothing was, where nothing could have continued to be, and that of the enormity of death.[3]

Defamiliarization is a term coined in 1917 by literary theorist Victor Schlovsky to describe the way art makes us see things anew. Such a process can stretch from a mere redescription to a life-changing moment.

> Defamiliarization is, then, both a function of art and a structural feature of spiritual life, and whether or not it is recognized in this way, it is the source of deep resonance between them . . . it celebrates the epiphanic quality art can have when it strips us of our limited views, of what we think we know.[4]

Henry Shukman describes thus how art is a pre-eminent way to free our seeing of the habitual. It exists to restore us to raw experience, freed from 'what we think we know. . . . When we deeply engage with good literature, good art, it changes our customary view of things. The world itself seems different, clearer, closer.'[5]

The path is that of attention. In the words of Guy de Maupassant:

> What you have to do is to look at what you wish to express long enough and with enough attention to discover an aspect of it that has never been seen or described by anyone before. There is something unexplored in everything, because we

have grown used to letting our eyes be conditioned by the memory of what others have thought before us about whatever we are looking at . . . To describe a blazing fire and a tree on a plain, we must stay put in front of that fire and that tree until for us they no longer resemble any other tree or any other fire. That is the way you will become original.[6]

Two hundred and more years earlier, the Japanese poet Basho exhorted us 'from the pine tree, learn of the pine tree, likewise from the bamboo'. Such attention, such learning, is alive, creative, vital and cleanses the doors of perception from memory and habit and what T. S. Eliot called 'the knowledge that imposes a pattern'. Such knowledge, such second-hand patterning, prevents us from seeing that every moment may be a new pattern; one that challenges us to respond freshly. Jane Hirshfield is one of the finest guides through the processes of poetic attention, As poet and Zen practitioner she is well aware of the linkage between attentiveness and creativity, suggesting that the writing of poetry must be thought of as much as a contemplative practice as a communicative one. Its practice, she says, 'requires an equally intensified listening, as a violinist must listen to orchestra, violin and body, if he or she is to play well'.[7] Of art and meditation she has written:

Art is one way a person can choose to enter, choose to fully know the range of human existence and experience. There are other ways. Zen meditation practice is one. Both are paths of awareness that allow us to move inside our own feelings.[8]

Such attentiveness to experience inner and outer also enables connection with what is other than the self: 'Art is a way to release our attention from immediacy's grip into gestures that encompass, draw from, and remind of more expansive constellations and connections.'[9] She sees the work of artists as transformation,

transformation that comes from a careful direction of attention, desiring

to perceive the extraordinary within the ordinary by changing not the world, but the eyes that look . . . It may feel as if we have done nothing, only given a little time and space of attention; but some hairline-narrow crack opens in the self's sense of purpose and there art, there beauty is.[10]

A work of art, she has written, 'is not color knifed or brushed onto a canvas, not shaped rock or fired clay, a vibrating cello string, black ink on a page', but is the result of a changed attention, which she describes as 'our participatory, agile, and responsive collaboration with those forms, colors, symbols and sounds'.[11] Our attention is the heart of the matter, able to change the mundane into art.

Awareness, whether in practice or art, asks a question:

'What is worth paying attention to right now?' That could be my personal life. It could also be some larger question, shared by all. The questions of political intransigence, partisanship, and violence; the questions of the unfolding environmental catastrophe we are living within are things that my poems turn toward, as much as any more individual sorrow or question. Awareness is always the starting place. Awareness shows us the questions, the problems we might be able to solve and the questions that can't be answered at all, and awareness makes the hand-holds and toe-holds appear, as we traverse the cliff of our lives. It also makes the cliff appear, and the lives, and the hands . . .

Closely attended, any moment is boundless and always changing. You emerge from these kinds of undoing awareness and you know it is not you yourself who are all-important. You know something of the notes of your own scale.[12]

Such attention to experience, whether arising out of internal thought and imagination or from external perception out in the world, when the self is left behind or no longer inhabited as the centrepiece of the experience, opens us up to new ways of seeing, as another commentator of spirituality and landscape describes:

> seeing reality as it truly is. Doing that requires being present to the moment apart from the expectations and interpretations I bring to it. Once I stop shaping reality into a theatrical performance with myself at its center, mindfulness allows the world to surprise me. The universe becomes delightfully open-ended . . . The mindfulness that wild terrain evoked is actually a sort of 'mindlessness', an end-run around rational analysis that seeks an immediacy of presence.[13]

6
Attentiveness to the Word

A writer, I think is someone who pays attention to the world.

Writers have a paradoxical task. It is language, at least ordinary language, that alienates us from the unfolding ever-changing process of life, yet for writers, language is their medium of expression. Language (particularly the English language), the foundation of imagination and narrative, supports the illusion of the separate and permanent self. It is enshrined in grammar. It structures our mind and our thoughts so thoroughly that we are mostly not even aware of its shaping power. Only close attention to how we use language and how our minds work can reveal this and offer alternative strategies. It is, perhaps, not surprising that Marcel Proust understood deeply this work of the writer to undermine habit:

> This work of the artist, this struggle to discern beneath matter, beneath experience, beneath words, something that is different from them, is a process exactly the reverse of that which, in those everyday lives which we live with our gaze averted from ourselves is at every moment being accomplished by vanity and passion and the intellect, and habit too, when they smother our true impressions, so as entirely to conceal them from us beneath a whole heap of verbal concepts and practical goals which we falsely call life . . . Our vanity, our passions, our spirit of imitation our abstract of intelligence, our habits have long been at work, and it is the task of art to undo this work of theirs . . .[2]

Martin Heidegger also understood this, writing in several places that language is the house of being; that language distinguishes that realm where humans can meaningfully dwell. Yet he sought a transformation of our use of language, a transformation concerned with openness to Being. He proposed a distinction between ordinary *speech* and what he termed *Saying*, where *Saying* is perhaps nearer to listening than to speaking, entailing a receptivity rather than mastery, a letting appear, perhaps another way of describing the collaboration of which many writers speak. 'Man acts,' he said, 'as though he were the shaper and master of language, while in fact language remains the master of man.'[3] Heidegger's call for humility in the face of language and experience, and his anxiety that today we deal with representations and dead speech rather than immediate experience, is reflected, though from a different angle, in the writing of Matthew Crawford discussed earlier. In fact Heidegger himself used the example of a cabinetmaker's apprentice as one who 'makes himself answer and respond above all to the different kinds of wood and to the shapes slumbering within wood . . . Without that relatedness, the craft will never be anything but empty busywork.'[4]

In contrast to our usual idea that we are the masters of language, Heidegger was concerned with letting language itself speak, writing an essay, 'The Nature of Language', intended to bring us face to face with a possibility of 'undergoing an experience with language', where to undergo an experience

> means that this something befalls us, strikes us, comes over us, overwhelms and transforms us. When we talk of 'undergoing' an experience, we mean specifically that the experience is not of our own making; to undergo here means that we endure it, suffer it, receive it as it strikes us and submit to it.[5]

This repeats the idea that such a conscious experience with language is transformative. As humans, he believed, we belong within

Saying: 'The essential being of language is *Saying* as Showing.'[6]
'*Saying* that shows makes the way for language to reach human
speaking.'[7] In our fallen or everyday manner of speech, far from
dwelling in language within *Saying* as the house of being, we use
speech, which merely designates. In many different ways artists of
the word attempt to transform language to express this enriched
dimension, revealed through enhanced attention, to the world
without and within and to the language with which to express it.

Language and Land

Wild land in its immensity or in its specificity assists attention in
opening up a canvas that is wider than our concerns, putting self
into perspective. Thus not surprisingly, many of the writers that
to me appear the most attentive write of the natural world. As
Native American writer Linda Hogan says: 'There is a way that
nature speaks, that land speaks. Most of the time we are simply
not patient enough, quiet enough, to pay attention to the story.'[8]

If we can find that patience, listen quietly with close attention,
something changes. Henry Miller described it: 'The moment one
gives close attention to anything, even a blade of grass, it becomes
a mysterious, awesome, indescribably magnificent world in itself.'[9]
Such attention does not always demand novelty: but it does
require our time. The novelty is in the looking, not in the object.
As earlier quoted, Wendell Berry writes of the purified attention
that finds newness even in the entirely familiar. He says that 'to
know imaginatively is to know intimately, particularly, precisely,
gratefully, reverently, and with affection'.[10] For years Cézanne
painted Mont St Victoire, never twice the same. Practices in attention, practices that encourage us to see what is there, and to know
it as it is, beyond the scope of our self-centred concerns, open up
such experiences into an unexpected richness. Virginia Woolf
described these as 'moments of being', and Nan Shepherd as the
'irradiation of the common'.[11]

Roger Deakin, who wrote wonderfully of wild places, wild swimming and wild woods, was described by another nature writer, Robert MacFarlane, as 'an explorer of the undiscovered country of the nearby'.[12] It is the attention of care that reveals the familiar as vivid and new, just as MacFarlane himself discovered. An account of his own journeys to find the wild places left in the British Isles ends, 'Wildness was here, too, a short mile south of the town in which I lived.'[13] In a later book MacFarlane quotes the Irish poet Patrick Kavanagh, who wrote: 'To know fully even one field or one land is a lifetime's experience. In the world of poetic experience it is depth that counts, not width.'[14]

MacFarlane has recently united attention to landscape and attention to language in a series of books, *The Wild Places*, *The Old Ways* and most recently *Landmarks*. Of another writer, Nan Shepherd, he has written that her 'sentences are born of patience and attention'.[15] The two for him are so combined that he believes that language deficit leads to attention deficit, which in turn leads to impoverishment of experience. As a bulwark against this, *Landmarks* contains pages of glossary, precise regional terms relating to features of landscape and weather and country practice, now close to extinction. All his writings are a hymn to the richness of embodied, embedded experience, a call for re-enchantment, a defence against loss. 'Language is fundamental to the possibility of re-wonderment,' he believes, a walking riposte to technological imagination.

Again with MacFarlane, as with so many writers and thinkers, we find a plea for humility and openness, a letting go of finality and certainty. He describes a 'decentred eye and a centreless nature' with which 'walking becomes a means to a certain kind of knowledge – one of the constituents of which is an awareness of ignorance'.[16] He writes that the 'true mark of long acquaintance with a single place is 'a readiness to accept uncertainty': a contentment with the knowledge that you must not seek complete knowledge. Such an appreciation of the relationship between things, land and watcher or writer, includes care, both in the Heideggerian and the more ordinary sense. It also includes tact: 'Tact as due attention, as in tenderness of encounter, as rightful tactility.'[17] We touch and in turn are touched in reciprocity of contact that gives life to the felt world.

MacFarlane embodies a listening that shifts ego focus from the centre to a place of participation, similar to one described by another writer he much appreciates, Nan Shepherd, who wrote that 'the focal point is everywhere. Nothing has reference to me, the looker.'[18] In another reference to the work of Shepherd, MacFarlane writes of attention as devotion. From this shifting from an egocentric perception to one of participation arises, almost organically, an appreciation of interconnectedness. Such collaborative participation is echoed through the work of all the writers above, from Basho's exhortation to learn of the pine from the pine through to a true embodied and experiential illustration of philosopher Merleau-Ponty's engagement with the 'flesh' of the world.

Attention and Imagination: The Fictional World

So many of these themes – defamiliarization, the evasion of habit, the refreshing of experience and the participative collaboration of writer and reader – arose again, in relation to fictional worlds of writing, when I spoke with novelist Ruth Ozeki, the author of *A Tale for the Time Being*, shortlisted for the UK Booker Prize in 2014. Ruth had exemplified for me the saying by Ortega y Gasset, quoted at the very start of this book: 'Tell me to what you pay attention and I will tell you who you are.' She inhabits the attention of her characters to such an extent, that we, her readers, are transported into their experience. When we met at a Buddhist study centre in Massachusetts, I asked Ruth, an ordained Zen Roshi as well as a writer, about the importance of attention for her, both as writer and meditator.[19]

'As a writer I have a sort of phenomenally long sense of attention, I can pay attention for very long periods of time which is good because of course as a novelist, one needs to. I seem to be able to, in the right conditions, work for really long stretches. It is a little contingent on where I am in the writing process – first drafts are always a little bit more difficult as you are generating your material and the way I write, I don't know where I'm going. So there is the sense of stepping out into the void and it's not a question of attention so much as the strain of that, the physical or mental stress of that is harder to sustain over a long period of time. When I'm doing revision, rewriting or editing, I can go for hours and hours and hours, and I have to stop myself. It becomes very important to set bells and limits. What happens is that I don't pay attention to anything but what I am paying attention to, and so I'm paying attention to the writing and I forget to pay attention to my body, the world, I forget about everything else, so I generally have to set some kind of alarms or bells to wake me up out of the dream.'

When I asked whether her Zen practice had instilled or helped this ability, she replied: 'It's certainly helped it. It's made me more aware of the importance of setting an intention and regulating, having a schedule of some kind. Zen is all about the schedule, the bells ring at certain hours and whatever it is that you are doing when the bell rings you lay it down and you move on to the next thing in the schedule. Transitions from one activity to another are done very quickly and efficiently in a monastic setting and that's a useful skill. I think that one of the problems that most people, and I certainly have, is making that transition from activity to another in life – it's just difficult and we resist transitions, we hang on to whatever it is that we are doing now, and so the monastic training is really useful in acquiring that clean movement from point to point during the course of a day.'

She told me that she regularly taught her creative writing students mindfulness exercises, saying, 'one of the things when I am teaching students meditation, part of what we are trying to do is to teach them to sustain the quality and duration of their attention in any given moment. But the other thing that I am trying to do is to teach them to come back – this idea that if, for example you are paying attention to your breath and then you stray, and then you notice that you have strayed, and then you come back. Norman Fisher [Zen Roshi and Ruth's teacher] talks about meditation as a process of return and I think this happens in the moment too, moment by moment, as we are sitting we are constantly straying and coming back, straying and coming back. It also happens when you are trying to start a regular Zen practice for example, a regular meditation practice. You miss a day and then it gets harder to come back. It's the idea that you let go of your attachment to doing it perfectly in the moment and then you just come back and start again. And breath counting is the perfect training for that – you lose count in the middle and then you come back. A lot of my students tend to be quite perfectionistic and so this idea that they can, quote, screw up, they lose count, to be able to let go of that without judgement and come back is a really useful skill. There are so many studies now concerning "grit", the idea that sticking with something is what determines success or failure in the end, more than raw talent or any of the other things. I think that's an important skill. Learning to sustain attention is learning to let go of your attachment to a kind of perfection – forgiving yourself for lapsing, for the lapses of attention and if you can do that quickly, if you can let go of your attachment to the perfection of your attention, then you are more likely to be able to sustain your attention, and then you can take that into your writing, you can take that into anything you do in your life. So my feeling with my students is that they may or may not become writers but if I can give them a little bit of very basic mindfulness skills, then that seems to me to be helpful.'

Such mindfulness exercises are also used more directly in relation to writing fiction, in service to entering the attentional world of the characters. Ruth explained: 'In writing we talk about it as perspective or point of view and this is something else I work with my students all the time. We do these mindfulness exercises where we will sit and do a body scan and just do basic mindfulness and then we will work on five-sense awareness, or rather six-sense awareness, thoughts being included. [Buddhist psychology posits six senses, mental sense being the sixth.] Then what I ask them to do, I ask them to remember that feeling of being alert and embodied in the present moment now and then take that feeling and with your imagination put yourself inside your character. Sit there and do a body scan in the body of your character and then open your eyes in your imagination and look around, see what your character sees, see what they are paying attention to, see what they notice, what they are feeling with all of their senses and then just make a note of all the things that the character has within their sensory field and then allow yourself to just follow, to go in any of those directions. That's what you're doing there; you're paying attention from the point of view of your character. It's all an imaginary exercise; you are sitting there with your eyes closed but your character leads you further into whatever world he or she is inhabiting.

'I think the non-judgemental stance of meditation is very important. Chekhov talks about how one of the important things is not to judge your characters; that your job as a writer, I can't remember his words, but your job as a writer is to write the truth of that character and as the writer not to judge that character. I think many writers have said similar things, so if the character for example, as in the reading that I did today, suddenly starts spinning off into this fantasy about love hotels, I just have to pay attention to what she is writing and not get self-conscious about it, or edit it out or anything like that; just let it emerge, let it bubble up. Then later, there are always sober second thoughts. You can go in and exercise quite a bit of editorial control. It's a negotiation. In the first draft things are often overwritten, so I go back in and tone it down, but the essential voice of the character you have to respect. You have to pay attention to it, you have to let the character do or say what he or she wants to do.'

Our talk then turned to *The Face*, an ebook that Ruth had recently published, and she described the process. 'In *The Face* I sat down to do this three-hour observation exercise, sitting in front of a mirror for three hours, and making a note, making a time code of any thoughts or observations that I had as I was sitting there watching, observing my face. I had just read this article by Jennifer Roberts from *Harvard* magazine about the power of patience and also the power of immersive attention and I knew I had to write this essay because I had a commission. The idea of writing an essay about my face suddenly hit me: "What have I said yes to? What have I agreed to do?" But after reading the Jennifer Roberts article, I thought this could be a nice structuring device for the essay, and so I then had to sit down and do it. It was an interesting experiment because it was very difficult to do at first. My eyes just wouldn't stay fixed on my face, they wanted to slide off in one direction or another and then I got locked into a staring contest with myself and stopped observing anything interesting, so I had to force myself to, then go through the parts of my face and really direct my gaze at different parts. Then once I had this time code done, I let it sit for quite a while, maybe even six months or so, and went back when I could read it fresh and I looked at it and just free associated. Anything that I saw in the time code that might suggest another side essay or slightly divergent path, I would make notes about that. Then I had this spine. I had never written anything like it before and I probably never will again, but it was a very interesting exercise. I think when I had been writing *A Tale for the Time Being*, I had been in the process of trying to write a memoirish kind of thing: when fiction writing wasn't going well, I would shift over and I thought it would be a memoir about approaching ordination and taking care of my mother who had Alzheimer's. That was the idea of it – and then in the end what happened was that the space opened up in the book for the Ruth character to step in and I realized that it's kind of perfect because here I am as a novelist trying to write a memoir and not doing a good job of it; failing to write this memoir, and then I thought how does a novelist, a fiction writer, fail to write a memoir? Well, she turns it into fiction and so the memoir in *A Tale for the Time Being* is Ruth's failed memoir. Having done that, writing the *Face* essay was interesting to me because I had already been playing around with personal narrative and the introspective gaze and so being able to couple it with this kind of strange observation exercise which I really was thinking of that too as a variation on the charnel ground meditation in the *Satipatthana Sutta* in the sense that if our greatest fear is death and decay then the best way to confront that fear is not to look away, is to look at it. At the age of 59, now I am sixty, looking in the mirror is that, it's a practice of looking at decay. I was very aware of that and it was a very, very interesting exercise. It was hard in many ways and a little bit emotional. Not terribly emotional. I think I was really holding back in certain ways. I wasn't letting myself feel a lot. It was only when I went back to it later that I was able to feel the emotions that were lurking below the surface of the text.

'What happened afterwards was that I went outside, after three hours of sitting staring at myself, I bought a latte at the corner coffee shop and I went to Thompkins Square Park in New York and it was a beautiful spring day and I was outside and the sun was shining, the trees were bursting with leaves and blossoms and there were people just enjoying the sunshine and walking around and I just remember sitting there on a bench and just looking at all of these people with faces. They all had faces, and I suddenly realized that each one of these people has the same kind of conflicted relationship with their face that I have with mine because that's the nature of faces. I had this real sense of this kind of swell of compassion, because we see faces all the time, we don't look past the skin, we don't look past it to imagine "what is that that man sees in the mirror in the morning? Who does he see? Does he see his father? Does he see his mother? Is he happy with his face? What does he hate about it? What does he not mind about it? What does he even like about it?" All of these things. "What makes him look away? When he sees that big mole on his nose does he avert his gaze, or does he not even see it?" Suddenly all these faces became this incredible terrain for curiosity and inquiry.'

I told Ruth how, after reading the essay, I had practised this kind of defamiliarization myself on the New York subway, and how, doing so, the passengers became more human, became alive in a different way. I suggested this was perhaps analogous to the way she inhabits her characters in a novel, that whether they were real or fictional the process was similar. She agreed that this was actually very true. 'I have always had a problem making that distinction between the fictional and the real, especially when it comes to writing. Anything that you write is fictional. The act of writing it down fictionalizes it. Then it really is a contractual matter between the writer and the reader. If I say this is non-fiction and I give it to you and you read it with that understanding of my intent, but also understanding that it's non-fiction but that doesn't necessarily mean that it's true in any kind of absolute way. The way I see writing is that it's always a participatory, relational art form. I feel that a novel is a collaboration. I do my part to the best of my ability and then I put it out to the world and there is this object there, and it feels as if it's not changing, it's real, but at the same time everyone who picks it up and reads it interacts with the text and reimagines it and in collaboration with me creates a completely different world. I don't know what that world is but the novel that you and I read is going to be completely different from the one that someone else and I made together. And I won't know. We all think we are talking about the same book when we talk about it but we are not.'

I suggested that fiction writing has always asked for a quality of attention from the reader in a way that perhaps visual art has not. Ruth agreed: 'It's one of the reasons that I got out of filmmaking. Because I enjoy the collaboration that words can evoke and I'm more comfortable with it. Film always felt two things to me; it felt a little bit bullying. When I used to edit film, I was very keenly aware of the power of music and image, for example, to manipulate emotions – hugely powerful. It was fun to do but it was almost too easy to do. There wasn't the latitude there. Whereas with writing you are giving over more to the reader. The reader takes on more responsibility for co-creating that world. The reader has more agency. The reader can read when she wants to, put it down when she wants to. Of course you can turn off a video too, turn off the television or whatever, but you usually don't read a novel in one sitting either. It expands over a longer period of time. And I think that what I have learned – I have worked in text, in film and in audio – visual motion picture imagery simply isn't good at a certain kind of complexity. It's very good at sensual complexity, mood and tone – very good at those nuances of feeling, sensual response. Aural sound or words without images can handle a little more complexity, but it's still a time-based medium. What the written word does well is that it handles this enormous amount of philo-sophical complexity, and you can do both at the same time. And this is what I am always trying to do. I'm trying to play with these thematic complexities and melding worlds together and layering, I am constantly layering levels of meaning and so I am playing with that, but at the same time trying to tell a linear story that moves quickly and efficiently through time, because you don't want to bore your reader either. So it's a question of two things: can you move diachronically and can you create a complex synchronicity happening at the same time?'

I suggested that this would be completely impossible without really well-honed attentional skills. Ruth agreed: 'That's true and it takes a long time to do. It means you are constantly going back to the beginning and you are pulling strands through, layers of meaning, symbols, words, sounds, you are just constantly going back and pulling these layers through, these threads through – coming back, coming back, just as in meditation. I remember get-ting intrigued with this, I think it was when I was reading Blake in college and I was thinking a lot about voice and irony, the structures of irony, and reflexivity in writing and layering of imagery. So I look back and see that I have been interested in this for a very long time. I am always playing with the reader's expectation, because to me it is a conversation. I'm not really thinking about the reader when I'm writing, because I don't know who the reader is going to be, but there's a sense of playing with the world, playing with the world's response and I think on some level I always assume that my reader is a very good dear, intimate friend who's just a little bit flirty in a nice way. There is always this playful element, "Let's delight each other, let's play." It's not didactic. It's done in a sense of "We are doing this together, let's have fun." That goes back to Buddhism again. Listening to what I have just said: that attitude towards the world, that attitude towards others, comes from intimacy; it comes from seeing no separation. You're not making assumptions about your reader as being very other from you; your reader is an intimate friend who is very close to you. I have been travelling with this book for three years now and I have met thousands of readers in many different countries. Even with translation something has survived and I do think that what-ever it is is coming from practice, an attitude that I have developed through practice, and received from my teachers. The person who is very present in my mind, whose voice and attitude I hear is Norman [Fisher].'

While *A Tale for the Time Being* is imbued with the writing and the outlook of the thirteenth-century Japanese writer and teacher Dogen, it still speaks as resonantly for those who have never heard of Dogen or cared about Buddhism, which I think comes from the way that Ruth gets inside her characters, into their attention, which takes us, the readers, immediately into their world. Ruth said that she did think this really came from her practice of trying to imagine herself inside the skin and the mind of her characters. She quoted Azar Nafisi, the author of *Reading Lolita in Tehran,* as defending the writing of fiction from precisely this point of view; that novels are inherently democratic because they give voice to different points of view, and they also develop the 'empathy muscle'. She even suggests that teaching novels is a defence against fascism.

Poetic Attentiveness

Poetry, with language and form most concise, most intense, most formal, perhaps illustrates verbal attention most nearly. It can encompass, often in short form, what is said and what points beyond saying in the white space between the words – the dialectic of sayable and unsayable, emptiness and form. As Emily Dickinson wrote: 'I dwell in possibility –/ A fairer House than Prose –/ More numerous of Windows –/ Superior – for Doors.'[20]

The very devices of poetry – metaphor, simile, image – lead us beyond the literal, into double or multiple meanings, into paradox and expansion, which increase our attention and amplify our experience in reading them. When I first contacted Jane Hirshfield requesting a meeting, she replied that 'attention is very dear to me'. When we finally met in her quiet house and garden in the shadow of Mount Tamalpais, north of San Francisco, I asked her to expand on this. She told me: 'As far as I can say it attention is the unifying love of everything I do in this world. What is writing except a way to attend through a particular set of available instruments, the instruments that language offers of meaning, of association, of reference, of what can be called in through language, of the music of language of the shapes – because with poetry the shape is so important? These are all tools of creating a score for an awareness that cannot be reached any other way. If you love attention, if basking in attention is what makes you happy, as it does me, then to find new shapes, new forms, expanded modes of attention, this is just an increase in happiness and an increase in knowledge, compassion, existence, interconnection. I don't see the way you can get to any of those things without first attending. It's difficult for me to understand, except when I am very tired, the choice made by so many people in the contemporary world to surrender their attention to the shabby and the superficial. I do understand it, there are times when to dunk your mind is all you can do in your exhaustion and I think that is one of the prices we pay for the busyness of the lives so many of us lead. But it is not a joy: it is a rest. When I look at what I want for any day, what I want is to be aware, and what I seek are the modes of entering awareness more deeply than if you aren't amplifying it. So art is one such thing, meditation is one such thing. But for me one of the lessons of my eight years of full time Zen practice is that if meditation were something of the zafu, the meditation cushion alone, it would be worthless. The point of sitting in concentration is not to find that an end point but to find in it an instruction for what you are looking for all the time. If it's not with you when you are not on the cushion, to whom does it do any good? Not you, not the world. But to bring that awareness into daily life and to find in the ordinary blue cup (pointing to the blue mug on the table) a reminder of all the possibilities of awareness, that's what meditation teaches us. It teaches us the vulnerability that allows attention in. It teaches us that the permeable is not frightening.'[21] This reminded me of something she had written: 'Poems do not simply express. They make, they find, they sound (in both meanings of that word) things undiscoverable by other means.'[22]

I asked her if the meditation or the poetry had come first. 'The poetry started in childhood. As soon as I was taught to write, I took to writing. All my childhood I wrote, often for the audience of the mattress. I hid things. It was not for display, it was not because I wished to be seen. It was because it was a place to explore the world in great safety and great privacy, where I could explore the world without risk, or risk without penalty. So that was always there. But oddly enough when I look at the story of how I arrived at the gate of the monastery in 1974, looking to see what might be inside, it was through literature. The first book I ever bought for myself when I was eight years old was a $1 book on Japanese haiku. When I think of that now, having learned a thing or two about haiku, I wonder what was this eight-year-old girl seeing in this that she so loved. I was living in New York City and this was natural world poetry that drew me and yet I was magnetized. So right from the beginning of my life in poems, there was work, which basically came out of a Buddhist background, leading me in, serving as a gateway. Everything that I loved fits well with Buddhist practice, Wordsworth is actually deeply Buddhist [she refers to a book by Robert Blyth on Zen and English Literature]. All of those things I loved. Even when I was studying Latin what I loved were the Epicurean poets who are very close to Buddhism in what they are saying and feeling. "You have this moment. This is the moment in which your life is enacted." So I always loved things that led towards things which I later discovered you could describe as the teachings of Buddhism.'

This expression of poetry as a contemplative art, this intensified listening, is the very heart of Jane's poetry. I asked her about a statement she had made about poetry coming into being through the 'fracture of knowledge', and about uncertainty. She has written, 'Inclusion of the impossible, the unsayable, and paradox is some part of how the enlargement art brings us is made.'[23] That enlargement comes not from answers or from certainty but from what escapes them. Just as centuries earlier John Keats had written of what he called 'negative capability' – 'that is when a man is capable of being in uncertainties, Mysteries, doubts, without any irritable reaching after fact and reason' – she has stated that literature's, and particularly poetry's, work is to find a way to live with and alongside the uncertain.[24]

Laughing, she said to me: 'I have long felt, and I love it that I say this with great certainty, which is its own undoing, that certainty is one of the most dangerous of human conditions. If you are sure of yourself, you are closed to all other concepts and if you are sure of yourself you can do terrible things out of the certainty that you are correct, whereas to always question is to throw open the gates to being corrected. It's the antidote to our arrogance and our hubris, which is one of the things we humans are so susceptible to – all those Greek tragedies are teaching us the penalty of certainty, of hubris, of clinging to a narrow and strict interpretation. I heard when I was starting to practise and it has stayed with me all my life, that Buddhist practice rests on a stool of three legs; great effort, great faith and great doubt.'

We spoke of the difficulty of remaining in uncertainty: the almost inevitable loss of self-reflectivity and self-reflexivity of institutions, even those founded on ideas of impermanence and emptiness, and the danger of even the most radical doubt turning ironically into certainty. Jane felt that we are all susceptible to this, 'which is why we must remain uncertain of ourselves. And to take this over to the practice of poetry, one of the marvellous things about writing a poem is that the poem is written and in the next moment you know – first you find the draft which is full itself of moments of probing and unknowing, and waiting and having something slip in through the silence of not knowing what will the next line say, what will the next word say, wondering "what will the next music be? What on earth will the ending be?" I have never written anything where I knew what the ending would be. Once I did and then it was very difficult to turn it into a poem. It was a prose statement that I knew had a poem in it, but I had never written a poem knowing beforehand where the ending was going to be, and it took three months to allow enough uncertainty in to open it up, to have it become a poem and not an idea. In writing a poem, one thing that happens, for me, is when the first draft has somehow been arrived at, the next thing that you do, is you read it over with a set of questions. What else? What's different? Do I really feel this? Have I left something out? Have I put too much in? Is it alive? So the whole process of revising for me is to be uncertain about what it is that I have set down and by that uncertainty, the revising can help it return to the first-draft-mind

– become something better. There is the first thought, best thought idea, but not for me. For me sometimes. Sometimes I might ask all those questions and not change a word, or merely change a comma, but lots of times, first thought, starting thought, and then only by allowing the uncertainty back in can the clay become malleable again. Awareness and attention are one way to bring the suppleness of existence back into the room.'

When I suggested that it is through uncertainty that the door opens in to a more participatory understanding, Jane replied, 'Art is always participatory. Words on a page are dust without the attention of a reader. Art always takes place in us and by our response. It's how it moves; otherwise it's electrons spinning around. Uncertainty might simply be described as opening your eyes. One of the most quoted things I have ever said basically encapsulates all Buddhism in seven words: Everything changes. Everything is connected. Pay attention.

'I do think that, even though it leaves out a lot – it doesn't have the word compassion in it. But I think if you pay attention and you notice that everything is connected compassion is an inevitable response. And so uncertainty is the open window. It is the oddest and most marvellous paradox that for me in my life – I talk about my life and I say me, and yet the experiences that I treasure most are the ones in which "I" vanish in which I have fallen into a condition in which the ego is not there. And what do I want most? I want to vanish into everything but not be dead.

'That's the big self – you know, that marvellous big self, small self. The self isn't lost; it's just not constricted into a clenched fist. And when we talk about holding something, what is held in an open hand is not gripped, is not clung to. It's held by collaboration and balance and agreement by both parties. And maybe this is how we want to hold our lives, not with closed hands but with open hands. Attention is always – the kind we are talking about – there are other kinds of attention and they are good too, but the softness of attention . . .'

She went on to tell me of a recent experience with an unfamiliar form of creative art with which she had experimented while staying at an artists' colony: 'Because I was a novice I was using "hard attention" because I was afraid of hitting my hand, missing, making a mistake, and I could feel how this was not useful, and how it was a novice's error of attention to be focusing so hard on making sure that everything was in the right place. It was a terrible expenditure of extra effort whereas when the studio assistant was doing it, for him it was easy because he trusted his hands and I was terrified of my hands. I could see that with more time I would get past it and that's so often the case, and I think it's important to have compassion towards the times when we grip and the times we are frightened and the times when our attention is the wrong kind of attention and we know it and we still can't change it. The process can't be willed. It's a funny collaboration of intention and attention. Intention is so important. It is the way you can invite without force. Intention, like attention, they can't be gripped too hard. And they allow for change. Another word that goes well there is invitation. I am not able to change my poems by willpower but I can change them by invitation. The muses are not very obedient to orders – at least my muse isn't but she will sometimes accept an invitation.'

Invitation appeared to link back to uncertainty and openness and participation, and I suggested that some of the reward of that invitation, of that attention, might be an enhanced appreciation, a change in the eyes that look to reveal the ordinary (the blue cup), in a new light, as epiphany, the 'everyday sublime'. Jane told me a story of a meeting she had had some years back with the poet Robert Bly, who had asked her if she was really referring to an 'ordinary white bowl' as exemplifying the sacred. She had told him: '"Actually that's exactly what I am trying to say because if it's continuous and interconnected you can't separate out the ordinary white bowl and I'm a great deal more comfortable with it myself than with gods and angels." He actually heard me and he wrote something later that showed he had got it and respected it.'

This transformation of the ordinary, by opening up a space around it, is a step towards the enlargement that Steiner described. Jane noted: 'We go to art as we go to a library, to a church; places where people fall quiet are places where something larger can come into us.' The something larger enables the coming into being and the presence of non-being, and death, that 'startles its reader out of the general trance, enlarging a bearable reality by means of close-paid attention to attention's own ground'.[25] She illustrates this process with a poem by Izume Shikibu she had translated many years ago, which says that awakening, often signified in Japanese poetry with images of the moon, will only be achieved by allowing and experiencing the full range of events and feelings:

> Although the wind
> blows terribly here,
> moonlight
> also leaks between the roof planks
> of this ruined house.[26]

It is only through the undefended, 'ruined' gaps in the ceiling that the moon can enter. Jane remarked: 'For me that was a life-changing poem. For that poem I had the words, I had the grammar, I knew there was something really great in there but I couldn't quite figure out what it was saying, and until I understood it I couldn't translate it. Finally the meaning came through and the poem changed my relationship to everything once I understood it. That was 1985, and it simply changed my relationship to difficulty and hardship – the understanding that if you want a human life you must be willing – you can't only say give me the moonlight, because if you wall off the moonlight, you wall off everything. If you wall off the pain, you wall off the moon. You must be permeable. So it was a poem which actually altered my relationship to being a human being and living and going through this world. I'm very glad you quoted it and so then it can help somebody else. Once I could translate it then other people could get that meaning.'

Poetry, she has written, 'comes into being by the fracture of knowing and sureness – it begins not in understanding but in a willing, undefended meeting with whatever arrives'.[27] With its clarity and abstention from clutter, the haiku form expresses this most evidently. Haikus are small jewels of open attention. Of one of the most famous haiku poets, Basho, celebrated also for his *haibun*, which combine travel journals with poems. Hirshfield wrote that he 'concerned himself less with destination than with the quality of the traveler's attention'.[28] With this fineness of attention, a good haiku points, as Hirshfield explains, both to world and to self, yet to a self that is transcended, unfixed, enlarged by its implication in world. Another Californian poet, steeped in Zen Buddhism, Gary Snyder explains this journey beyond self-expression into presence, writing that: 'every haiku is a selfless poem. The only way a haiku can exist is by a momentary act of such selfless clarity that there is no way for the ego to mess it up.'[29] As Hirshfield says: 'They unfasten the mind from any single or absolute story, unshackle us from the clumsy dividing of world into subjective and objective, self and other, illness and blossom, freedom and capture.'[30] Art, she states, is never sufficient in itself. 'The ground of any artwork's existence is a human life, psyche, mind and heart, and the transformations in them it awakens,'[31] the transformations both of artist and reader or recipient.

TO RECEIVE WHATEVER arrives, to be transformed, for poet and for listener, requires attentiveness and openness. With attention and openness, poetry catches the nuance of the moment, the radiation that transforms the common and expresses it in most crystalline form. It is, I think, in poetry that language comes closest to being disclosive, performative. As Paul Celan wrote, 'In a poem what is real *happens*.'[32] Heidegger wrote, 'Thought and poesy are in themselves the originary, the essential, and therefore also the final speech that language speaks through the mouth of man.'[33] In his later works Heidegger frequently turned to poetry to express *Saying* (which perhaps is not so far from what Jane Hirshfield called 'unsayable') in contrast to the ordinary language of *speech*, an understanding of *poesis*, that is, making.

He wrote that 'Projective saying is poetry . . . Poetry is the saying of the unconcealedness of what is.' Furthermore, he felt that 'The nature of art is poetry' and even '*All art*, as the letting happen of the advent of the truth of what is, is, as such, *essentially poetry*.'[34] Writing himself in poetic form, he said: 'But poetry that thinks is in truth/ The topology of Being.'[35]

While he states that poetry in the narrow sense is only one mode of 'the lighting projection of truth i.e. of poetic composition in this wider sense', he also considers that 'the linguistic work, the poem in the narrower sense, has a privileged position in the domain of the arts'. This is because language, as the house of Being, brings beings into the Open for the first time by naming them, and only thus are beings brought to word and appearance. All other creation depends on this and can happen only in 'the Open of saying and naming'.[36] And this projective saying brings forth not only the sayable but simultaneously the unsayable into the world.

As discussed earlier, writers have a paradoxical task. Their medium, language, at once alienates them from the interdependent process of life, supporting the illusion of the separate self, that in English at least, enshrined in the grammar, unthinkingly structures our thoughts. One of the greatest sources of alternative strategy as shown above is found in East Asian poetry. 'Around the seventh century,' says David Hinton, 'Chinese poets began stripping poems down to the bare essentials. They were pushing at the edges of language, trying to make language point to something outside itself. I've described this before as "voicing silence".'[37] In his fascinating and poetic book *Hunger Mountain*, Hinton shows how the Chinese language is better adapted to expressing this silence, and he makes one of the most successful attempts to convey this to his own experience in Western terms. From a Taoist perspective, echoing many of the themes expressed by Hirshfield and Snyder, Hinton describes the way spiritually engaged Chinese poets used poems as a form of practice to get past

> the relentless industry of self, each thought and feeling and memory appearing out of emptiness, wandering through various transformations, and disappearing back into emptiness, I see that I am fundamentally separate from the mental processes with which we normally identify instead, I am most essentially the emptiness that watches thought coming and going, hears leaves clattering down through bare branches.[38]

In a most thoughtful conversation with David Hinton, he told me that he felt the Chinese poets he has translated express simply what Heidegger was stumbling towards, deprived of the concept and deep understanding of the silence and emptiness that was pivotal to their perception. I asked how important attention was to him, and he responded: 'My first thought is to change the word to attentiveness. It connects with emptiness. Meditation is attention to attentiveness I guess. The first thing you notice is your thoughts are going on and you are watching them with attentiveness and you realize that you are not your thoughts. The first thing that attentiveness shows is that you are not that analytical rational machine and if you keep watching then your thoughts get quieter and quieter and then you start noticing that thoughts appear out of nowhere and disappear into nowhere. In Chinese thought the cosmos was made up of absence and presence – two fundamental principles – Buddhist words for emptiness are synonymous with words for absence, so absence is emptiness in Buddhist thought. The ten thousand things emerge out of it and they go back into it, so presence is the ten thousand things, the empirical universe, so when you are sitting you see thoughts coming out of absence.

The next thing that attentiveness shows you is that consciousness is absolutely part of the same tissue as everything else, as all
of empirical reality; there is no fundamental difference between
subjective and objective processes. We generally assume that we
are somehow separate from the world. Why we think that is a long
story . . . but then the last thing that you notice is if you sit long
enough, if you meditate long enough, you see that your thoughts
slowly might fall silent and there are no thoughts and then there
is just the absence, the emptiness. And from there, that is the
beginning of all Chinese art because from there you realize if you
open your eyes there is no content in consciousness and you look
like it's mind as mirror and it mirrors whatever you look at exactly.
Again there is no difference between inside and outside. So that's
attentiveness at its most intense. And that is the deepest spiritual
place to be because then you are the cosmos looking out at itself
and there is no distinction between – there is no you there anymore
– there is just this cosmos attending to itself and that becomes really
fundamental in Chan or Zen Buddhism. And in the arts where, if
you look at landscape paintings, you are supposed to look at them
that way, with empty mind mirroring and when you look into one
of those paintings you see the paintings full of emptiness, empty
space and landscape. There is absence and presence. I think that's
the heart of attentiveness in ancient China and it still works, it's
empirical observations; that's how consciousness works, that's
how things work.'[39]

He compared this deep approach to attentiveness favourably
in contrast to the contemporary fascination with mindfulness for
stress relief and so on, feeling that such self-serving intentions
behind the practice 'miss the boat'. On the other hand 'the old
Taoists and Chan Buddhists, and artists and poets were not interested in happiness, they were interested in some kind of deep
dwelling, and their attentiveness is the door to that deep dwelling.
Dwelling is part of that cosmological process of absence and presence because in attentiveness you actually become part of that

process, you dwell as part of it.' This kind of deep attentiveness, this openness, he felt, is also that of the poet: 'that level where you notice that thoughts come out of nothing and go back into nothing, that's the attentiveness I think, that we might say a poet has that. In a sense, all thoughts are the other, it's only an illusion that somehow we are them, because they just go on by themselves, it's almost the cosmos thinking in us. So if you start paying attention to that, it's a little like what a poet does, it's just for a poet it might be a little more interesting or turn into a poem.'

He believes that the Taoist and Buddhist understanding of emptiness and interdependence allows for this attentiveness to open up beyond the ego but without losing touch with the world in a manner closed to Western Christian beliefs. 'Attentiveness is the way out of identity, what people call ego. It's an opening that you experience yourself as part of everything else if you know what you are seeing. In the West, mystics, or those who may have experienced something like this opening, have a whole conceptual structure within which they interpret it as god's love or god's grace or union. It's an experience; I think the way that I have described it, the way the ancient Chinese would have described it, is as an empirical description of what's going on whereas Christian mystics have laid over it this whole story over this core experience though so they didn't really know what silence was, what emptiness was because they immediately interpreted it through this whole conceptual complex.' All these thoughts expressing the Chinese conceptual structure are expressed in Hinton's most recent book, *Existence*, which was published after our conversation. In it, he limpidly reveals the foundation of Chinese thought and aesthetics through descriptions of one scroll painting and its attendant calligraphic poem.[40]

This way of attending and experiencing challenges the Western tradition in a radical way. It is interesting that it has certainly influenced several of the best Western poets. Gary Snyder, Jane Hirshfield, Mary Oliver, Hinton himself and many others have been influenced by this different style of looking at the world. Imbued with deep attention and understanding of the intertwining of emptiness and form, absence and presence, self and other, it sustains poetic expression as dwelling or presencing. Such a foundation perhaps enables the deepest, most non-dual way of responding to Octavio Paz's statement that

> All poets in the moments, long or short, of poetry, if they are really poets, hear the *other* voice. It is their own, someone else's, no one else's, no one's and everyone's. Nothing distinguishes a poet from other men and women but those moments – rare yet frequent – in which being themselves, they are other.[41]

While such work, aware of both ground and particularity, is perhaps more easily accessed from the philosophical, artistic and practising background of an Eastern sensibility, some Western writers and artists have also expressed such understanding clearly. William Carlos Williams's famous red wheelbarrow stands out, along with many other, often American, examples of stripped-bare attention. Little known but highly respected American poet William Bronk brings a Western responsiveness to this form in many short exposed poems; in his case mostly expositions of fact or thought rather than object, in which, as in Japanese and Chinese poems, the foundation is attention. Where Rilke said that the work of the poet was to praise, Bronk, who also says, 'I praise' in a poem entitled 'Praise the Music', even more fundamentally says the poet's duty is to attend, even when articulacy fails, and what is is almost incommunicable. Even when 'Awareness listens, tries, can't say',[42] yet we attend.[43] And in the attention, despite all his doubts about world and self and meaning and his often painful struggle with meaning, 'The earth is beautiful beyond all change.' And his attention is rewarded with epiphany, an epiphany of a tree in the middle of a field: 'This tree! This tree!'[44]

Writing of poetry in an essay, he says:

> Poetry is about reality, the way that a lens is about light . . .
> It is best when it is clear and transparent, when it is least
> there, in the sense of calling the least attention to itself . . .
> The lens of poetry gathers from many sources, draws upon
> many aspects of what there is . . . It makes nothing, it changes
> nothing, but it focuses on reality, on what there is, and it
> illuminates and clarifies. . . . One might say, for trial, that
> poetry is a statement about what there is, so attentive, so
> scrupulous, that it partakes of the nature of its subject: what
> there is, is poetry; it is not made; it is attended to.

Bronk continues:

> Is it not also true that it is the nature of what there is,
> closely attended to, that it cries out for the directest kind
> of statement? It is my conviction and practice that this cry,
> these statements, are poetry also, impossibly so, but so
> nevertheless.[45]

The work of the poet is hard, repetitious and unknowable
but, he says, it leaves the poet with a job to do. And that job is to
attend. Bronk later changed the name of this essay, 'The Lens of
Poetry', to 'The Attendant'.[46]

Another American poet, Jack Gilbert, echoes the Chinese both
directly, tempted by Chinese poets and their immaculate pain on
seeing the reflection of the moon in a bucket, and in many other
short and long poems, songs of hearts that are 'lost in dark woods'
yet always suffused with a hard kind of joy. Despite pain, 'what
astonishes is the singing'. Like Bronk, he attends to the world,
with attention, a kind of astonishment and a negotiation between
despair and delight. A characteristic poem entitled 'The Answer'
begins with a question: 'Is the clarity, the simplicity, an arriving/ or

an emptying out?' And there is no answer, only a memory of walking through a Greek village where someone was playing Chopin.[47] At the end of joy and despair there remains, as he called one of his collections, 'the dance most of all'. Encompassing attention inward and outward, feeling and intensified form, these poets achieve the enlargement and transformation that Hirshfield points to.

The stakes are high. Paul Celan would seem to have been searching to express this same kind of open attentiveness in his *Speech on the Occasion of Receiving the Literature Prize of the Free Hanseatic City of Bremen* in 1958, when he suggested that a poem as a manifestation of language and thus essentially a dialogue can be a message in a bottle, sent out in the belief that it can be making its way towards something, 'toward something standing open, occupiable, perhaps toward an addressable Thou, toward an addressable reality'. Employing the word translated into English as attend, and using all the meanings of the word, he stated that he believed ways of thoughts like this

> attend not only my own efforts, but those of other lyric poets . . . They are the efforts of someone who, overarced by stars that are human handiwork, and who, shelterless in this till now undreamt-of sense and thus most uncannily in the open, goes with his very being to language, stricken by and seeking reality.[48]

In a later speech at Bremen, now known as 'The Meridian', he speaks of the mystery of an encounter: 'the attentiveness a poem devotes to all it encounters, with its sharper sense of detail, outline, structure, colour, but also of "quiverings" and "intimations".' He then quotes the saying by Malebranche that occurs in Walker Benjamin's essay on Kafka, 'Attentiveness is the natural prayer of the soul'. Such attentiveness makes a way, opens a meridian, enables encounter with an *other*, even he suggests a *'wholly Other'*.[49] Where perhaps those poets coming from the Eastern traditions would find the other not separate from self, same and difference, empty and full – not one, not two – non-dual.

MANY OF THE THEMES encountered above, different forms of attention, the constrictions of self, the importance of not knowing, and the centrality embodiment, were echoed in a fascinating conversation that explored the intricacies of writerly attention and creativity that I had with Alice and Peter Oswald. Alice is one of England's leading poets, winner of the T. S. Eliot Award and the first Ted Hughes Award. A review of her latest collection, *Falling Awake*, published after the date of this conversation, described her as 'fierce in the quality of her attention'.[50] Her husband Peter, a playwright of distinction, is the author of many verse plays and translations and a former playwright in residence for the Shakespeare Globe Theatre in London. When I spoke with them not far from the River Dart, the subject of Alice's best-loved long poem, it became clear that they both made a distinction between two forms of attention. The first, which they called focus, would seem to coincide with what most people would think of *as* attention – our conscious orientation to a task. What was obviously more central to both of them, which alone they named as attention, was what I would called that creative or more broad-ranging attention expressed above, that somehow bypasses the willing self. As Peter explained: 'To me focus is central and visual, what is strictly in front of you – and attention is about the periphery and listening to the periphery, which does require an emptying out of the centre. Focus intensifies the centre, and what's going on in the periphery, maybe non-human things: attention is an all-embracing thing. In focus you have a name but with attention you have to empty yourself out and see what's there. In writing, it goes in waves, there's all kinds of intention and short-term aims – to finish a line or a scene or whatever which requires focus, but then for me, it's almost like wearing out the mind, the focus almost exhausts the mind to the point where it faints and goes into a kind of knocked-out state which then switches over to attention which requires an absence of volition. Maybe that's why fasting and shamanic practices create a kind of knocked-outness, which is not just unconscious but kind of purposeful in a paradoxical way, and in that state new content comes in from the periphery where you haven't been looking – or, as a playwright, voices that you are trying to listen to. You can crowd them out with your intention but then, in that slightly knocked-out state, you can maybe hear them again and maybe the play or scene or line goes off in a different direction. I was also thinking of soldiers standing to attention. There is a similarity there, because you are looking straightforward but not focusing on anything.'[51]

Alice agreed and turned the conversation to the central importance of rhythm, thus linking verbal and musical attention, and starting a fascinating conversation between the two of them about the process of writing. 'I think that's incredibly brilliantly put and I agree with Peter. To me the interesting part is the paradox that you are somehow there and not there, and that's the struggle with any artistic work I think. You have to eliminate yourself and yet remain attentive. I am interested in the technicality of that and particularly how rhythm can achieve that. I suppose my particular interest has always been in those oral traditions where there were years and years of training. Given the hexameter, they were given this extraordinary long line that has a sort of wobbling effect on the mind. It puts your physical being into a precarious state that actually creates – not attention but tension, because it has this way of flowing where it's changing its patterns all the time. It starts on a very strong beat unlike a pentameter. This extraordinary line that the Greeks came up with has a sort of physical effect on the body. It makes you slightly off balance so you are alert and physically tense, and yet at the same time you have got this machine going on – the lines keep coming and coming and the rhythm is sort of the same. So you're made anonymous by that mechanical effect but you're kept there and alert by the overbalancing effect. So I'm particularly interested in how poetry, and the oral tradition in particular, maintains that paradox and keeps something vacillating between absence and presence, which I think is absolutely what Peter was driving at. You've got that movement between focus and attention, the knocked-out and presence, and intelligence and stupidity that you need for writing. And for me, and I'm sure the other arts have different ways of doing it, for me, it's entirely to do with the patterning of rhythm. It's about creative repetition. You are repeating but in such a way with such variations, that you have to keep on top of it.'

Peter added that he thought 'there is a similarity between the line ending and the end of the out breath, when there's a pause between the out breath and the in breath, and into that something new can come. Similarly at the end of a pentameter or a hexameter, there's just like a pause and then the next one starts. There's something analogous to the widening of attention in that little gap which is a bit like the wobble Alice is talking about, a kind of swaying like a dance.'

Alice then suggested that she thought that's what verse offers beyond prose: 'It's so intensely physical. It is such a dance of the tongue. I am interested in this word 'stupidity' because you need a certain amount of stupidity to let something else come in. If you are too intelligent with what you are doing, it's a kind of defence against the rest of the world. As Peter says, what you are really wanting is new input. You want to damp yourself down so that something you haven't thought of comes in.'

Peter related this to translation. 'There's that distinction between attention and focus. There's the text in another language that you are focusing on and there's your own version that you're writing and also focusing on, but somewhere between those two there is that point of concentration where the mind sort of faints – and into that comes not just the repetition of the verse, but something that is new but obviously related to the original. It's that gap. It requires intelligence but not limitless intelligence so that at some point the intelligence will just let go.'

To Alice this was reminiscent of what she had experienced when she was writing *Memorial*: 'When I did my first run-through of it, I was looking really hard at the Greek. I had all my dictionaries and different versions and I was comparing everything and writing lots of notes. And then I lost the notebook that I was doing all my work in. I left it in a hotel in Cornwall. I was completely devastated. It was about a year's work. So then I had to start again basically. I thought "I really can't go through all of that again, so I'm going to get rid of the dictionaries, get rid of everything," and I was aware, physically aware of a sensation like a solar eclipse. It was as if when I had been using all that paper and pens, my own body had actually been moving in front of the text and casting a shadow. The thing you are very aware of with Homer is that it's extraordinarily full of light and suddenly when I abandoned the notebook and the paper I was aware of moving off the sun, so that something else that was nothing to do with me, was able to hit the page, which was a really interesting experience.'

Memorial, subtitled 'An Excavation of the Iliad', based on Homer's *Iliad*, is Alice's most recently published work. In another interview, and resonant with the idea of poetry as performative, she has said of Homer: 'He just transmits life. No mediation. He describes a leaf and you don't get a transcription of a leaf, you get a proper leaf.'[52] She has also said elsewhere that she doesn't put pen to paper until the poem has formed in her mind, saying both that she likes the body to take part in writing and also that 'It's a question of trying to take down by dictation what's already there. I'm not making something, I'm trying to hear it.'[53]

I asked them if they had ways of practising attention; practices that act as resources for them. Alice's answer was definitive: 'All the time. To me my whole life is about maintaining or creating that level of attention. Everything that I do I regard as a preparation or invocation for work. I suppose that because for a long time I worked as a gardener, I did develop a technique of preparation for writing. It wouldn't be that you were thinking of a poem, it would be – it's hard to define exactly – but it's just a way of filling yourself to a point of readiness. I don't even know what it is that I'm doing, but I know when I'm doing it. It's particular doing repetitive tasks. At the moment I'm working in a plant nursery on weekends where they have a potting machine and I discovered it's lovely just to shift pots because something happens in your mind. I don't really know what it is but you can turn almost anything, except, I find, cooking, into preparation.'

Peter described how he used to have elaborate rituals – 'specific points of accumulated energy, like a tree I would walk around – or standing still in one spot for an hour or two. In London it was on the roof and someone once saw me and shouted out asking if I was all right, obviously afraid I might be contemplating jumping. Now I do a more concentrated sort of meditation – like drawing down a certain kind of energy, but it certainly isn't to do with focus. Walking – walking down the lane where we live there are many flowers and you can stop and look very carefully at one flower or you can just walk on with banks of flowers like Ladies Bedstraw on either side. Attention for me is that second experience, where you can't name or even draw as you don't truly see it, you're not looking at it, there's just these waves of white at the periphery. Something completely different is happening and that energy is very uplifting for work for me – a kind of filling up.'

Reiterating the way certain themes keep reappearing in relation to attention in so many different arenas, Peter echoes the mechanic and writer Matthew Crawford in bringing up the idea of submission: 'I think attention is a kind of submission, obeisance. Certainly with acting. Not having had a great training but being thrown into it with my company I depended on peripheral beings. Michael Chekhov describes how, when he left Russia for Germany and had to learn a part in German and perform almost immediately on arrival, he had this split experience where he would watch himself performing from the periphery of the theatre.[54] A similar thing happens with me, allowing or creating – I don't know which – a bright sphere up in the periphery, attention to which creates the energy in me, which enables me to transcend my normal state and perform with focus. So the attention then leads into the focus. What is required is not exactly belief but I guess faith and trust in these peripheral beings.'

Alice described how, 'If I've got to recite something long, I make a kind of triangle of imaginary beings around myself. So I'll put something in front and two things behind myself. At the moment it's been a grasshopper in front and two goddesses behind. It's like this kind of wedge. I feel protected in that I'm attending to them. If I look at the audience in a real way I will be scattered, so if I've got this prow or ship I'm protected.' And so the discussion returns to the body, the importance of embodiment, and Alice describes her preparation for attention: 'When I actually sit down to write as opposed to preparing for writing, it's very physical for me. So I will ensure that I'm very symmetrical yet slightly off balance too – in a symmetrical way. I find if I'm too much slumped over or my legs crossed, it's like the energy is in some way obstructed. I have to set myself in order to be physically receptive.'

She describes meeting a Japanese calligrapher who had insisted on the importance of correct posture in a manner unusual to Western sensibility. 'It's incredible how the body can teach the mind. You think you're stuck, then you simply change posture, or relax your shoulders and you move on.'

For Peter, one of the greatest forms of attention is walking, simply walking and noticing what happens to your mind: 'Specially if you've been focusing, which obviously uses your body and mind in a completely different way. And then you just go out and walk. I also find that before performing my long story poems, I learn the lines outside in the fields. If I'm standing out in a field or wood or whatever just reciting these lines in iambic pentameter for a long time then my surroundings completely change. I think it's tuning into the periphery, or again an emptying out of thoughts because there's not room for thoughts as well as the recitation. It's not like that gives me a clearer sense of myself, it gives me a clearer sense of my surroundings, which I think is attention.'

And so we go back to definitions and to what is happening in the brain, the neuroscience of default attention and focused attention, focused attention and the wandering mind, narrative focus and the present-moment global attention, Iain McGilchrist and the methods of the two hemispheres. What was perhaps missing in the neuroscientific descriptions and theories but what we have been talking about is the working of the creative mind. What Alice and Peter have illuminated is the feel and the rich experience behind the description. The next day I got an email from Peter, which brought in another aspect of attention, its capacity for more than singular focus, its shared aspect:

Dear Gay, it was lovely chatting to you yesterday about attention. One thing occurred to me later, thinking of Arthur Miller in a note in *The Crucible* saying of a character who refuses to lie, 'Attention, attention must be paid to this man.' If he had said, 'Focus on this man', then it would feel like he was addressing himself to the reader as an individual; but saying 'Attention' feels more like addressing a group, or people generally. So it occurred to me that attention has a non-individual quality and that, if you pay attention, or enter into attention, you are entering into a field where the boundaries between individualities are dissolved; one effect of this is that you yourself, the attender, are attended to, so there is a sense that by giving attention you are also receiving it, and from innumerable beings, in a way that's hugely healing. Miller is calling for the whole world to pay attention to this man because the man is paying attention to the truth. Well, I guess the thoughts this subject invokes are endless ... better stop there! Love from Peter.

Endless indeed.

7
Visual Attention

When I think of art I think of beauty. Beauty is the mystery of life. It is
just not in the eye. It is in the mind. It is our positive response to life.
AGNES MARTIN[1]

The discussion with the Oswalds revealed a clear distinction
in their minds between what they called attention and what
they thought of as focus. Attention was, for them, the creative
state. This is somewhat similar to the two types of attention that
Paul Klee wrote about as practised by the artist. He thought that
the normal type of attention focuses either on the positive figure
enclosed by a line or, with more difficulty, the negative shape that
the line cuts from the ground. The good artist, however, according
to Klee, will be able to hold all the elements of a picture in a single
undivided attentiveness, aware of both the obvious positive and the
less usual negative aspects; a scattered, scanning, undifferentiated
attention that holds the total being of the work in a single view.
Reading this, I remembered that when I started my earlier book
on emptiness, my artist friends, in reply to my question as to what
emptiness denoted for them, said that it was space or opportunity,
rather than being merely a lack, which was the answer of most
non-artistic contacts. Indeed, in relation to Klee's second form of
attention, art psychologist Anton Ehrenzweig wrote of the 'full
emptiness' of the unconscious scanning that appears in the work
of painters and other creative workers.[2] He considers it unconscious, stating that the gestalt principle ruling conscious perception
is unable to let go of its hold on the figure, which we consciously
see as either positive or negative as displayed in the duck/rabbit
or the old woman/vase configurations. Today it seems to me that

it may follow from McGilchrist's careful distinctions between the two hemispheric modes of attention, relating more to right- than to left-hemisphere predominance, which of course reflects the over-widespread unscientific relating of creative work to the right hemisphere. Ehrenzweig felt this 'full emptiness of attention' also exists in hearing, following Klee in describing this form of undivided wide attention as both 'multi-dimensional' and 'polyphonic'.

It is the task of the great artist to integrate these two forms of attention, uniting this multi-dimensional scanning awareness with their conscious focus and intention in the work. Painters have always drawn our, their audience's, attention through their use of light, illuminating what they wish us to attend to, to the point where today, for James Turrell, light is both subject and medium. Standing in front of the great Caravaggio painting of the *Beheading of St John* in the Co-Cathedral in Valetta, my attention was inescapably directed to the illuminated areas. First, the main subject, the just-completed murder – the bent back of the murderer, the woman holding the bowl and the shoulders and face of the fallen St John. Then it was directed, by light again, to the second subject, the pitiless watchers – the two faces at the prison window above and to the side and to the second man above the body. Across the room is Caravaggio's *St Jerome Writing*. Here the dramatic lighting draws our attention to the arm and hand with the pen, to the skull on the table and then secondarily to the face, old, scholarly and pious. In both cases the highlights demanded by the narrative are held in a formal balance of the totality of the painting.

If we look at art frequently and carefully, the way we have been invited and instructed to see may then be passed on into the way we perceive the world. As I write, I look out of my window on to a hillside that to me appears coloured in broad strokes like a Cézanne painting of Mont St Victoire. I see this in all weathers just as he depicted his mountain, time and again, in all weathers and all light. The attention that became creation has, years later and thousands of miles distant, changed the way I see a hillside this morning here on the west coast of America. Perhaps Cézanne is a great example of the rewards of training our attention. He was fanatical about it. If, after early development, deep changes to our neurological structure start from attention grabbed by novelty, or by willed practice, he may be a pre-eminent example of the latter, day by day, taking himself out to sit in front of *le motif*. Indeed his death from pneumonia came after even exposure to stormy weather failed to drive him indoors, away from his creative depiction of the mountain that obsessed him.

The Impressionists altered the way we see landscape and objects. In each generation, artists employ new ways of expression to explore our visual and cognitive relationship with the world. After the Impressionists, the motif itself loosened its moorings in the 'real world'. Through Cubism to abstraction via Surrealism, the perspective and the emotions of the artist and the viewer became of more import than any object of representation. Narrative and representation gave way to evocation, emotion and reflexivity as the viewer's awareness of their own perception became increasingly central.

It is a dialogue of attention between painter and viewer, just as the novel, as described by Ruth Ozeki, is a collaboration between writer and reader. Abstract Expressionist Ad Reinhardt wrote that 'abstract painting will react to you if you react to it. It will give you back what you bring to it.'[3] While for some what they could bring elicited sufficient reward, for others perhaps the demand was too much. The search of artists to enable their viewers to share their vision is ever changing. Today one artist, sculptor Antony Gormley, has stated that he has returned to the body as he feels that the vision of those pioneers of modernity (and he cites Mondrian, Malevich, Tatlin, Kandinsky and Brancusi), to create a language of formal purity that could touch everyone irrespective of race or creed, was a failure. Such failure, he suggests, came about because 'it was inadequately common and inadequately linked to experience.' Hence his return to the body, our common experience; yet it is not a return in the traditional representational manner. He writes of a 'transition from body as representation to body as space [as] . . . a translation from representation to reflexivity'.[4] His desire today is to bring back the body, not in a representative but in a reflexive manner, that will bring the viewer's attention back to their own embodiment in space and time. Art now, he says, 'is no longer about power and privilege, but participation'.[5] His work, he states, 'is rooted in a particular example of a human experience of embodiment. It is offered back to the world as a displacement, hopefully with some affordance.'[6]

The art object or process is a carrier of – that word again – affordance. The viewer now is increasingly invited to share the feelings, the place of the artist. Rather than a story or an object, what is on offer is an experience. Much contemporary art calls the audience to an *experience*. We are invited to share the attentive encounter of the creator. Painter Sean Scully has said that the artist is not able to change the world, but they may be able to change the viewer's relationship with the world.[7]

Take Your Time, said Olafur Eliasson in the title of an exhibition; feel how different your body/mind is when bathed in pink light from its immersion in green. Watch the rainbows as lights catch the water spraying up from your footfall.[8] Such pieces ask for a kind of contemplative meta-attention: attention to the very experience of attending. With light, as with attention itself, awareness commonly focuses on what is illuminated, not on the process of light or attention itself.

No one has perhaps asked us to change our relationship with the world more convincingly than James Turrell, who exposes and reveals the mechanisms of our perception in ways that disturb our conventional expectation. Turrell asks for this kind of attention through sculpting with light itself, rather than merely using light to reveal. 'The world is not a predetermined given set of facts; we construct the world with our observations,' he states, inviting us to share in his constructions – which will of course, because they are then ours, differ from his.[9] Like so many of the other artists

for whom attention is central, his work involves both deconstruction and construction and above all defamiliarization to induce a
novel experience. One of his earliest patrons, Count Guiseppe
Panza de Biume, said of an early work, *The Mendota Stoppages*,
that 'Turrell was able to change in his art . . . what is in front of
us, but which we don't recognize because habit makes it difficult.'[10]
As critic Andrew Graham-Dixon has described, he

> encourage[s] us all to pay more attention to the effects of
> light and air and atmosphere constantly at play in the world
> around us. His work can be intensely beautiful, intensely
> mysterious, but the work is only the beginning. How that
> work is absorbed by the consciousness, how it shapes and
> changes habits of seeing – that is what is all-important.[11]

This alteration of consciousness comes through a direct experience
of perception itself, stripped of narrative and image, a slow and
care-full attention that is close to meditation; an orientation prior
to discrimination.

> The deceptively simple forms of Turrell's art can channel
> the most subtle of experiences. Isolating and shaping the
> phenomena of light in space and time through our percep-
> tion, his art collapses the distance between the perceiving
> subject and the object of perception – akin to the Buddhist
> meditative practice of merging outside and inside to promote
> receptivity to a more spiritual, universal nature.[12]

In this sense Turrell takes us into a primordial openness, a notion
of visibility prior to distinction into inner and outer or emptiness
and form that is, to me, reminiscent of the Buddha's exhortation
in the *Bahiya Sutra*, that for the awakened person: 'In the seeing,
there will be just the seeing . . .' Indeed it has been described both
as 'a disorientating fullness of emptiness', and something that

the viewer 'no longer perceives but only sees', deprived of the orientation to which the mind is accustomed.[13]

Acts of closure or privation paradoxically enable openness to the presence of light itself. Curator Miwon Kwon has described Turrell's work as anti-illusionistic, saying that the carefully constructed architecture has to be made invisible 'in order for the seeming nothingness contained by them to be made visible as a presence'. The concealment of the means of production 'allows the emergence of light not as a metaphor or an illustration, but as a tactile material presence'.[14] Disorientation and privation that unhinge us from our usual moorings are obviously present in the works overtly titled *Ganzfeld*, in which the viewer is surrounded in empty space and boundaryless light, an experience similar to that felt in blizzards or flying through cloud, where all spatial orientation is lost. This experience has again been described as 'seeing but no longer perceiving', since the brain has nothing to latch on to, with which to orient or to make sense. It might be likened to the first stage of attention, 'alerting', which, as Rick Hanson pointed out, is a function of brain areas close to those that are activated when we experience 'oceanic feelings', or even to Keats's state of negative capability and unknowing. This sensation is also to be experienced in relation to some of the colour field paintings of Mark Rothko, Barnett Newman and Ad Reinhardt. Composer John Luther Adams has written of this *Ganzfeld* experience:

> Immersed in pure color, the viewer loses all sense of distance and direction. I long for a similar experience in music. I want to find that timeless place where we listen without memory or expectation, lost in the immeasurable space of tones.[15]

Indeed, Alex Ross describes Webern as writing ecstatically of being lost in a snowstorm on a hiking trip and walking into whiteness, 'like a completely undifferentiated screen', and suggests that his music offers a similar experience for the ears.[16]

Turrell's constructions sometimes literally take us back neuro-logically to the very foundations of sight. He is concerned with exposing and inviting us to share what he calls 'behind-the-eye' vision. In his *Perceptual Cells*, this may involve the viewer spend-ing time in a metal capsule, lying flat and isolated in a white space that is then filled with sound and colour. The intention behind this combination of intense colour with sound is to alter the brainwave frequencies of participants. One such work is *Bindu Shards*. *Bindu* refers to a Sanskrit sacred symbol of the cosmos in its unmanifest state, which is held to be the point that begins creation, where unity becomes manifest in the many. It is also a Tantric term used in connection with the energies of the subtle body.

Other works of Turrell's, his sky-watching spaces, involve the external natural world. These are sculpted spaces in which lighting and framing combine to intensify the experience of paying bare attention to the ever-changing sky. In these sky spaces we experi-ence the reversal of the usual conditions of viewing, the sky, no longer background but the active part of the exchange, looks back at us. 'The object of our vision, *habitually in front of* us, becomes the place of our seeing. We are *inside it*.'[17] Turrell has said:

> I'm interested in the quality of consciousness in space that occurs when you come into it, when you realize it's like an eye in the same way a camera has an eye, and space is somehow seeing and has a way of seeing made by subconsciousness. With my work you actually have a way to investigate this and look at it, but you do it by entering. It's like walking into the lens of a very large camera that has this special way of seeing. When we make the camera, it has its way of seeing that we tend to forget.[18]

As with Eliasson, we, the viewers, are required to slow down. Some of Turrell's works occur in the dark, and minutes of time must elapse before our eyes adapt to the darkness and the work reveals itself. For his most ambitious project he has spent the last thirty or more years actually sculpting an extinct volcano in Arizona. Passages tunnel through Roden Crater, exposing platforms and light shafts from which day or night can be experienced. Whatever the means, the central experience of Turrell's work is 'the emergence of light, not as metaphor or an illustration, but as a tactile material presence'.[19]

I spoke with James Turrell in Cornwall, at the opening of two of his works in the context of a wonderful garden, where contemporary sculpture meets extraordinary planting and landscape architecture in a hidden valley that once was part of a pilgrims' way. In his opening speech he had expressed his love of working outside, his appreciation of the English tradition of garden visiting and of follies, and his pleasure at the existence of his work in this place. This garden holds two works: *Aqua Oscura*, in an old underground water tank where long immersion in the darkness is necessary to see finally, as one's eyes adapt to the loss of light, the reflections of the canopy of trees above projected on the back wall above water; and *Ecliptic Elliptic*, an elliptical sky space originally temporarily installed to view a solar eclipse in 1999.

He agreed that his work does require attention: 'I don't know whether it's an intention or whether it's just a necessary by-product that is required. I simply like things to develop over time, like in the tank [*Aqua Oscura*] where you come in and you literally can't see anything.'[20]

When I suggested that while in the tank's darkness I had thought that the experience of many of his works takes us out of our normal comfort zones and puts us into a kind of limbo space in which, having nothing but expectation and waiting, you become unusually aware of the process of attention, he replied that 'this puts us into a comfort zone that is better than what we have been used to in this modern world. This modern world has got us out of a real comfort zone. I like it when you really do settle down and relax. I think it has to do with the combination of getting older and being OK with being older.'

I put forward the idea that in an increasingly secular world, the kind of heightened attention that his work elicits can become a form of spiritual experience. He replied that he thought that in the modern world, 'we have pushed things to the point where we have gotten so far from our comfort zone that we have thought that this is the normal way that things should be, or the way that they are, and this is quite artificial but I think that people feel relief to go into a situation where they are not worried about what it is that comes, where they have faith in it and it does come – and if you do deliver on that, then it works. And if you don't deliver on that, then they're not going to come back again, so you just forget it.'

He likened this to food: 'There's fast food and there's food that you prepare over a long time and it takes longer and there's some waiting and it takes people to do that and have some skill and then there is a reward. Sometimes the reward is rather subtle, a small thing. But we've forgotten about small things and how dear they can be.'

As to the way I believe his work elicits attention, he feels strongly that 'the attention is something that you give, not something that the object of perception essentially demands from you. If you don't give then, then it's over and you are out of there. It takes a curiosity that's over time and also perhaps that we develop as we get older. The young have no problem with this, they get right into it and then there are the old that do and it's kind of the in between that's having to actually make a living and do this world – and that's the sad part about it. There's a time I worried about that because you can be insulted about people who leave – or perhaps you can say that they are missing something. . . . Those are some of the thoughts that I have about attention. Now there are things that will create attention – things that do something that cause you to look and that's one thing. Different works that I do have different qualities in that regard. Some of them are just very missable – you can walk right by them, but others come out and get you a bit. That's OK. I am suspicious of myself in thinking about those things. But you see this in art a lot. A lot of it has more to do with the politics of how one lives in this world and things that might shock you. That's one thing. The other has to do with just attending to a phenomenon. A lot of people miss things – like a rainbow – this is a spectacular event and so many people are just driving around and they are not going to see it and there it is. So these kind of quiet things can be what I would say spectacular and over the top but if you are not looking, you don't see it.'

We don't see it, I proposed, because we tend to look in habitual ways. Turrell agreed: 'We have patterns of perceiving and we could call that prejudiced perception which it is, but we do get those patterns by the way we live, but the way we perceive is going to have a lot to do with what we can perceive, and I think art was meant to change that. That's one of its critical senses. I appreciate that and I'm very interested in how different artists do that.'

He said that he always wanted his work to be self-reflexive – that, when you see or look *at* the work, 'you think about how you look – about how perception takes place. Attending to a phenomenon is something that you award this phenomenon, that is, you give it – and so a lot of the things that we behold are things that we create. So not everything is out there for us to perceive: some of the things we create, and just like the context here (referring to the sky space), vision is what is changed. Obviously I don't change the colour of the sky but a lot of us are not aware of the fact that we give the sky its colour, so they don't understand how we do it. So sometimes I will change the colour of the sky – almost any colour you like – and people are quite shocked by this, because they don't realize that they are the ones that award the sky its colour in the context of vision. So they think its something that I'm doing to them. It's something they're doing, you're doing.'

To my suggestion that ideally, then, viewers will go away with a little shift in their perception, he said: 'Hopefully. It's a little gentle koan, a little gentle nudge. Not all work is for everyone

– that I have come to know. I have quite accommodated myself to that and that's fine.'[21]

Later that evening, more than fifty people filling the *Ecliptic Elliptic* sky space sat in spontaneous silence for 45 minutes as the dusk deepened into night, each one lost in the slow light show that unfolded in their perception around the egg-shaped hole in the dome that revealed the sky. The opening gradually darkened until it became opaque, a solid lid rather than a transparent aperture, as the plastered dome became translucent. Turrell is a magician who reveals truth: creating seeming illusion, he paradoxically reveals to us the creative illusion that is our normal vision. He reveals his true magic by deconstructing and disrupting ordinary perception, through inviting us to take the time to attend to its processes and showing us the illusion of our normal understanding – if we can award our attention. I doubt that anyone there that night will forget the experience.

Weeks later, in southern Spain, I slept in a room facing east where each morning the light through the window reflected onto the opposite wall a Turrellian artwork – a divided rectangle of light, apricot, green and gold, quite distinct from the light, white, over the mountain horizon and blue above, that appeared through the window itself. In the light of the Cornish experience and Buddhist reading, I enjoyed a multi-layered *appearance* of the coincidence of sense organ, sense object and sense consciousness that was creative and changeable, much richer than the usual uni-layered snapshot collapsed into a single named impression.

MANY OF THE WORDS spoken about Turrell's work could also apply to that of Garry Fabian Miller, an artist whose work is also centred upon light, though very different in scale and means, and concerned with the production of images rather than purely experiences. A talk with him echoed many of the themes already met in other conversations, other readings and viewings – light, openness, receptivity and epiphanies of the small and everyday. Miller is a photographer whose work comes out of a life of attention, and invites and instils attention in the viewer. His early work involved sequential images over time: one series illuminated processes of photosynthesis and another pictures of the changing horizon between land and water taken daily from his window on the Bristol estuary. Later, this mapping of sequential time accumulated into single abstract images made in the darkroom without a camera. His practice centres on a relatively narrow area around his home on Dartmoor, experienced in different weather, season and light, resulting in seemingly abstract images 'filtered through personal reflection and then recreated and interpreted as light events in the darkroom'.[22] He produces photographic prints made by channelling a beam of light through liquids, oil and water, in coloured glass containers producing patterns of light and colour that may be further manipulated by the use of cardboard stencils, discs, squares or rectangles that are sometimes pierced. Martin Barnes, the curator of photography at the Victoria and Albert Museum in London, has described how 'through this working process, the artist acts as a conduit: a bodily measure, an organ of sense perception and a psychological filter for the specificity of an environment.'[23]

When we met in Devon close to the River Dart and a little south of his familiar Dartmoor, I asked him directly how important attention was in his work, he replied (interestingly like several of the other people that I have talked with) that it was not something he gave a lot of thought to, but admitted that that would be typical of most things as he was unaccustomed to giving or writing his thoughts about his work. He suggested that some of the people I had mentioned talking to for this book were more in the system of explanation than he chose to be. 'I think I am interested in something called attention,' he said, 'but only because you have asked me. I wouldn't otherwise have given it a form.' Yet, I suggested, this was a gap between expression and actual practice, as it seemed to me that his work comes out of an enormous depth of embodied rather than articulated attention, which is (as Alice Oswald too had said) his life, the walks he does, the inhabiting of a fairly limited environment to great depth, which is a form of profound attention. He told me that Robert MacFarlane, in reviewing a book of his, *Home Dartmoor*, had described him as having a *chronic* relationship to this particular bit of landscape. 'Chronic doesn't sound a good word, but that's how it came across to him; that the need to be here, in this particular small space was so strong that he viewed it as that word. I found that a little troubling, as I would have said that it is a totally positive commitment to a particular place. If I stay within these boundaries whatever it is that I am seeking out will be exposed and I believe it's more likely to happen here than if I travel to all these other places. I have invested a number of years to believing this, so the investment of time is such that I think that whatever has happened so far has happened and what is likely to happen is more likely to happen because of what's gone before. So to walk away from here would be a mistake, but it's an act of faith to stay and believe that looking at the same things which are different every day; by looking at the same things, every day as more and more time passes you are going to get closer. The word I am interested in is exposure. I think that I am exposing myself to the possibility and I think that exposure is accumulation, in my case exposure is an absorption of light, time, all of which are important within a photographic practice. There is not too much, nor too little light – there is all the light that fluctuates across your day and your life and your exposure to it. I am committed to photography because I think it is a significant human invention, the way the materials work through the quantities of light touching a surface and reacting with it, causing things to exist which otherwise wouldn't exist. I think within my brain something similar to a photographic exposure is occurring. I remember some years ago – about 1994 – when I was doing an exhibition in Japan, which involved a lot of talking and workshops and lectures, I was having to think a lot. I thought at that time that the brain when you are born is completely absent of light, it was a dark space, and that your life was the exposure to light passing through your eye and then the quality of image, which was deposited or absorbed in your brain. So the exposure to what you saw, what you chose to look at, and how you chose to live, established the person you became. As a teenager, and then later in life, I came to understand that if I leave my enclosed space everything falls apart. If I stay in this space everything holds together. That's a pretty tough decision to take but I know if I stay in this space all the things that make sense hold together and within that framework the chances are that something significant will happen and if I go outside it things are going to fall apart. So that's the dynamic that I have accepted. I tend to think that is something that cannot be separated from the pictures I make; they go together. So people then have to give value to the method as well as the pictures, they become solidly linked and that's a rare thing I think in an artist's practice. Certainly for people who come from a photographic tradition it's a rare thing, it becomes a building block and you become a reference point for a certain view.' His words draw me back once again to the epitaph that opens this book, Ortega y Gasset's statement: 'Tell me to what you pay attention and I will tell you who you are.'[24]

Fabian Miller's experience and attention become embodied in the pictures, which then have a specific effect on the viewer. As he explained: 'Sometimes when I go to my exhibitions I can experience the response they receive from the work. My belief is that all the things that I am looking at that may apparently be invisible in the pictures – the energy which is in the pictures – comes from all the days of looking and things which I couldn't articulate but which I've absorbed, and hopefully they contain this unsaid thing. In Japan they would respond to the work and when talking to me would rub their chests just here [pointing], and they would talk about a kind of physical thing happening. You knew that these were ordinary people who were confident to say that to you, whereas people here would be lacking confidence to try and express the way that they feel in the presence of the pictures so would probably say nothing. If you take this small view of staying in one place, you begin to know that it works as year on year I am able to sell my pictures across the world. So this means that this very Dartmoor-centric view that I have is a global kind of multicultural experience, all coming from this place. That's the thing that probably gives me the most pleasure.'

I asked Garry about the possibility of this kind of heightened experience, energy and attention acting as some kind of a spiritual experience in an increasingly secular age, mentioning such artists as Olafur Eliasson and James Turrell. He distinguished his work from theirs in terms of its lesser scale and freedom from the need for installation and architecture, but also brought up another of the recurring themes of this book, the everyday sublime, the finding of the extraordinary within the ordinary or domestic. 'What is said about these artists is that words like spiritual are applied to them. So you are aware that they satisfy some kind of non-denominational view and need within people and that's important – and also the values that they embody are not really the values that the art world embodies so they are challenging to the art world, but they are the backbone of the history of art when art works properly and there is a very clear market for them. I don't think there is anything negative about this, though it becomes dangerous when the artwork takes a kind of position which would be viewed as idolatry, become sacred sites, spaces to which people have to travel to. I live an ordinary life and I want art to exist within an ordinary life and those transformative experiences, which you can experience this afternoon, just to be part of the fabric of your life. You can live with it in the ordinary way, you don't have to travel to these sacred buildings which contain special experiences. If you can go there, great; if you can't it has to be embodied in other ways. I would describe it as domestic. The opposite of rarefied and sacred, just within the moment of the everyday.'

He admitted that he cherished the romantic belief of what he termed the 'John Clare tradition': that there are many people, and he cited shepherds and lighthouse keepers, who understood such things and yet just lived ordinary lives. They experienced 'special experiences' that remained invisible, unsung. He is interested and admiring of the practice of potters as an ideal of the life of tradition and practice, citing potter Richard Batterham as a model of commitment to working and the sense of usefulness embodying an interlocking of making, function, purpose, physical process and way of life.[25] Yet he also felt that a more contemporarily ordinary life, exposing oneself through a diversity of places and practices, was not inferior, just different, and acknowledged that his life of commitment was unusual. 'There is a romantic idea about people like me because we are thought unusual. I'm not very comfortable with this but it's good that it exists because otherwise I don't think people will commit to thinking this could be a way to live a life and if I do I will find things which otherwise I will never find.'

I feel that this search for 'things which otherwise I will never find' is another of the recurring themes of deep artistic attention: a search for epiphany. Fabian Miller agreed: 'Definitely. Epiphany is key.' Earlier he had written: 'There's something I'm searching for and trying to discover. If I organize my life around these walks, there is a possibility that I might find it.'[26] So his life embodies this search for something; returning to the same places with a mind open enough to allow the unexpected to occur, exposing himself to the possibility that something extraordinary will happen there. He has pointed to an experience that Virginia Woolf wrote about in her diary of looking at the moon in the night sky and experiencing 'a great and astonishing sense of something there which is it'.[27] (In a memoir, Woolf writes of these 'separate moments of being' that are 'embedded in many more moments of 'non-being'.[28]) He explained: 'I'm interested in just the ordinary mundane domesticity of life, into which come the epiphanies in a kind of mind-blowing way and so that's how life should be.' He described an experience with his children: 'We are walking down an alleyway towards the sea in a built-up area and the light is hitting the sea in such abundance that there is a wall of white light coming down this alleyway and on this day I remember my younger children saying "Oh gosh an epiphany!" knowing that I was completely overdosing on it, and they knew that that's what it was all about and that's what I was trying to do and there it was just coming down an alleyway and they could make fun of me. It was just that, as awesome as that and then you just feed off that as long as you could carry it within you. It's that you were looking for. But you have to put the hours in. In my case it's the hours in the environment that I choose to believe can offer it, but it's also as an artist, in my case it's hours in the darkroom. For another artist it would be hours in the studio. I don't know if people still understand that: that notion of the artist painting, standing, physically engaging with the material, then perhaps abandoning the work of a day.'

He spoke admiringly of certain English figurative painters, of Frank Auerbach, Michael Andrews, Euan Uglow and 'the unbelievable nihilism of destroying every day's picture and starting again the next day, and then fifty years have passed. They embodied something about being an artist which is rare in the intensity of commitment and a belief that by looking at the figure for years with one person they would get to a place that they would otherwise never reach. That's pretty magical isn't it? It's an act of faith. I believe that if you put the time in and the attention, it's an act of faith whether the return will be there, but I think that my experience now is that it is so.'

While he much admires the work of James Turrell, in line with his emphasis on domesticity and ordinariness, for himself, he is committed to the idea of producing the kind of artwork 'that can exist without all the complexity. In my recent Edinburgh show at the opening, you were aware of a very immediate physical excitement of being in the space – the viewer was unclear what was happening, where all this light was coming from – it wasn't coming from a complex structure that had been created, it was coming from within the pictures . . . there was a real sense of excitement:

as more people came, the room built a sense of excitement at the physicality of the light and the way it was affecting them and the people they were talking to, the whole thing had an energy to it, which was speaking of itself – a really tangible example of an intense exposure. When I'm out in the landscape or when I'm in the darkroom very occasionally it's like that, but on the whole it's pretty steady and it's the kind of regularity and consistency and being there in all weathers and so on. It's the accumulation, which might just have some good moments, but on the whole it's just the day after day after day.'

He spoke of the approaching end to his practice as the supply of photographic paper essential to his dye destruction process of working will run out in the next year or two, and suggested that maybe under this pressure the pictures were getting more intense. 'As I get older I like them to be more active and more intense so that they give off the energy in a more immediate way. I think that some people don't like that, as the work is becoming more dramatic and highly charged; they prefer a quieter, stripped down kind of image. Perhaps I'm doing it because I think there is a need for it to be like that for people now. I think I want the effect to be rapid. I am choosing to make them like that, though that is not for them but for me as I feel there is less time left and I think I'm more interested in trying to get that – whatever that thing is – I want that to be very immediate and very active and to grab you in a very powerful way and hold you there. I think that's something to do with the end coming and I want to get to this place quicker than I might have done in the past when I felt I had more time. I think the end should be an optimum, maximum point. It shouldn't be just a quiet fading away. My attitude is, as it ends it should be an incredibly exciting sense of having arrived at that place, rather than just quietly accepting something and then fading away. Agnes Martin offers the other view, the quiet last breath, whereas I would like it to be an ecstatic moment before you blacked out. I think that the Agnes Martin view is one that people are a lot more comfortable with just now, that quieter restful last place. I have a lot of sympathy for the people who choose to live their lives in Northern Soul and rave culture, and those kinds of worlds. I have chosen not to be part of those worlds yet I think the people that inhabited those places were searching for something, that I feel very close to.'

Having completed more than sixty large-scale digital works that use the original Cibachrome print as the starting point, Fabian Miller has a body of exposures to work from in the future. This archive of exposures will be more than enough to last his life out, making only twelve to fifteen pictures each year. He is looking forward to curating his work, looking at it slowly over time rather than engaging with new work. This is the opposite of the attraction of novelty to grab our attention, closer to the intentional practices of attention, written about so well by Wendell Berry, who it turns out, was very important to Fabian Miller. 'He embodied the kind of thing we are talking about, the farming life and the writing that came out of it.'

Though Fabian Miller sees it as a romantic view and his illustration of shepherds, lighthouse keepers and domestic potters may seem of an earlier age, the number of books on nature writing, of attachment to place, even of the life of a shepherd, which, as I write, is in the best-seller lists, perhaps demonstrates the importance of such commitment to attention that is deep rather than widespread, even if it is commonly only embraced at second hand.

THOUGH HIS PHOTOGRAPHIC work is utterly different, Buddhist writer Stephen Batchelor is also a photographer and image-maker and echoes many of the themes that occurred in my conversation with Garry Fabian Miller. In an essay, 'Seeing the Light', from a volume on Buddhist influences in art, Batchelor mentions two themes that have constantly recurred in my journeys around emptiness and attention – attention itself and collage. His practice of making collages out of found objects led him to realize that photographs too are found objects; the raw data of the image is given, then 'moments in the visual field' like the pieces of a collage can be organized in the viewfinder to create an image. The attention and receptivity required for the good image are ways of cultivating awareness and transforming perception. He cites Roland Barthes in *Camera Lucida* making a link between the photograph and Buddhist ideas of emptiness and suchness in the way a photograph is a unique unrepeatable event capturing the contingency and evanescence of the moment. 'In order to designate reality, Buddhism says *sunya*, the empty; but better still: *tathata*, the fact of being this, of being thus, of being so.' Yet, casting aside specific Buddhist ideas as informing his practice directly, he says that photography, like meditation, is a tool for challenging his assumption that the world is just the way it appears at first sight.[29]

In a recent conversation, he said: 'Personally I have only found that the cultivation of attention in meditation feeds seamlessly into the use of attention in photography, collage or composition in writing. I don't think they are different things at all; the difference lies in the goals to which they are directed. It's a quite different intention when you utilize attention in making a collage or taking a photograph than when paying attention to your breath or your body or the sounds around you. It seems in a way to render something that in meditation at a certain level is relatively abstract into the production of a work.'[30]

He illustrates this by commenting on the way that the Buddha continually uses metaphors that compare meditation to the work of artisans. In the *Dhammapada*, he compares the fletcher who makes arrows, and the farmer who channels water, to the wise man who trains his mind. In the *Satipatthana Sutta*, the main teaching on meditation, he compares the person who, when he takes a long or short breath, knows that it is a long or short breath to a wood turner who knows when he makes a long or short turn and in another passage there is a comparison to a goldsmith. 'He must have closely attended to the work of artisans and that is an element that is not brought out in Buddhist theory as it turns into psychology defining what mindfulness is, and it becomes cut off from performance, whereas the Buddha seems to be drawing from his observation of artisans' work and how that is an appropriate metaphor for how we should understand meditation. I think that is rather important. Also in terms of what we have learned from biblical scholars is that the likelihood of a text being earlier rather than later is its utilization of parable and metaphor rather than argument. What happens as the tradition develops is that you lose concrete examples and it becomes more discursive, analytical and abstract. As we see in the *Abhidharma*, things are described in terms of what they are, no longer in terms of actual performative tasks. Such examples as the wood turner demonstrate explicitly that the act of paying attention is not about some purely subjective private qualities that are going on in your consciousness but about a relationship that you have with the world. It has to do with how you interface with your surroundings in order to achieve a specific goal. There I think it returns to what I was just saying about collage and art – it returns to the performance, it becomes performative.'

He sees photography as a practice of attention, and the camera as a force to see the world in a new way, as the lens is free from the editing that our minds impose, ordered as they are by our needs, aims and prejudices. He illustrates this in a series of photographs of reflections, showing how the camera gives equal attention to reflection in, say, a shop window, while the tendency of the everyday mind is to prioritize the object of our attention – the item in the window – and edit out the reflection, which is merely obscuring what we want to look at.

He also sees the way we configure what we see into a work of art as a transformation of objects in the world into something that is greater than the sum of what we see. In both his photographs and the materials he uses for collage, he selects those things and materials that are normally un- or under-valued and overlooked, bringing heightened attention to the ordinary and the mundane and seeing them in such a way that may transcend their ordinariness.

Speaking of this to me, he said: 'I notice that in collage work, part of the interest of working with found objects, and this was a deliberate choice, was to start paying attention to those things in our world that we habitually edit out, that we don't notice. And that, of course, is very similar to the formal practice of meditation where you pay attention to things you would normally not notice – your body, your sensations, your emotional states. And also when you go to one level higher still, to notice that these things are impermanent, *dukkha* and *anatta*. There is one passage [in the *suttas*] where I think it is Sariputra, is asked, "What is it that you pay *yonisomanasikara* to?" He says you pay embodied attention to the five aggregates *as* impermanent, unsatisfactory, not-self and empty (*dukkha, anatta, sunya*) and what is important is the *as*. It's not just eyeballing things; it's deliberately choosing to attend to features of those parts of your experience that you either deny or block out or ignore for whatever reason. And so likewise, with doing collage work or photography, you have to discipline yourself to notice what you would otherwise not see. And so for me there is a very explicit continuum between the cultivation of meditation in its Buddhist framework and the application of that, not just in the vague sense that they are both mind being attentive, it is a quite conscious choice to pay attention to certain things that we are not normally conscious of. And that I think is a very important part of art. Good art, for me, is not about producing something that is beautiful, it is producing something that enables or forces or shocks us to see the world in a new light. In that sense attention is about illumination, enlightenment in a way. And that of course is a very close parallel with Buddhist practice except that it is operating in a much more contained and concrete sphere of producing work. Practice and application.'

In an interview, he has written of deconstructing our habitual patterns of perception through stilling the mind, stopping the perpetual narrative, until something 'opens up into something not just more clear and vivid but also more beautiful'.[31] He describes his process of creating collages in terms of relationships: a physical relationship with the environment and with the body, the engagement of hand and body, texture and tactility with creative process, and the relationship of chance and order. Chance lies in the finding of material and order in the strict grid patterns of his organizing structure and the rules of composition in which no two identical pieces are allowed to be contiguous. These he relates to Buddhist *Madhyamaka* ideas of identity and difference. According to these, such categories are incapable of capturing the contingency, the fluidity, the emergence and the unfolding of life itself. His work then plays with the tensions between identity and difference in a manner that helps reveal the world as irreducible to such classical linguistic categories, 'which is *sunyata* or emptiness'.[32] In a recent essay he has referred to 'an aesthetics of emptiness' which 'originates in suspending belief in the inviolability of these distinctions [between identity and difference and self and other], while simultaneously highlighting the radiant contingency of life that is revealed once the grip of such dualistic thinking is loosened'.[33]

Such an aesthetics of emptiness relates his artistic practice to the field of Dharmic practice, uniting the ethical and the aesthetic. The link between them is the practice of attention that may reveal the world in a way that is refreshed, enriched and transformational, as the discarded, rejected elements of his collages are transformed into art works.

THE CONCERN WITH attention and physical engagement, hand and body, texture and tactility was reaffirmed in my conversation with Edmund de Waal. De Waal is a potter, an artist, yet is perhaps even more widely known as a writer, the author of *The Hare with Amber Eyes*, a best-selling memoir of a collection of beautiful objects (Japanese netsuke) and the lives and stories that collected around them. Today de Waal has taken pottery out of craft into the world of fine, even conceptual, art, with his installations of vitrines and shelves of still, quiet porcelain cylinders and bowls inserted into various environments from stately homes to galleries. I met with Edmund at his studio in south London, and it was rather like entering one of his works. A large, bare, hard-edged yet light-filled space, all white, only the occasional touch of grey, and a Bach chorale filling the air. We spoke first of his upcoming exhibition at the Royal Academy in London to accompany a book about white and porcelain. I think the distinction here is necessary: it is about far more than white porcelain. I had found that what Edmund has written and spoken about white resonated with what I had thought and written and discovered about emptiness. He had considered white as 'a staging post to look at the world from', as 'it forces other colours to reveal themselves'. As I had felt with emptiness, white leads on to attention as 'we look harder when we see white'.[34] He agreed with the closeness of his work to ideas of emptiness, saying that he defined himself 'as a person who doesn't make forms but makes emptiness or space, a series of spaces, of emptiness'. I had read earlier that he had said:

> I can absolutely remember as a kid the intense interest in the idea of making a space. It wasn't about what it looked like from the outside at all. It was that this idea of making an inner space seemed to me so extraordinary. And 44 years down the line I still get that – I still get that very strange bodily, somatic thing. It's very odd to make a volume in the world, to make a space, to define something. It's like capturing a bit of the world that isn't really there. It feels temporary, a brief holding.[35]

When I asked him about the place of attention in his work, he replied: 'I think it's pivotal in the sense that the thing that I focus on more and more is the need to slow down and to pause, and so one of the things that I try to do in my work is to create spaces and structures that have pauses in them which require attention. So when I am making a vitrine with lots and lots of pots there is also lots and lots of space in it, and so I'm hoping that the pauses and the gaps, the emptiness and the spaces and the vessels together create a duration, a kind of space of time to slow you down. Paying attention is absolutely critical and it's a skill that is not hugely present in current cultural life.'

He described how he had learned this skill of attention through his apprenticeship as a potter and also 'through a reasonably direct encounter with someone [Geoffrey Whiting] who was not only a potter but also a practising Buddhist, who also had come from a Quaker family, so had interesting disciplines behind him. He tried to work in silence where possible, and his clarity about the length of time it took to achieve – not only the mental time it took to acquire a skill ('the first ten thousand pots') – but the amount of attention you had to pay to what you were doing, was a very profound experience for me as a child. My choosing to be a potter very early on came out of this encounter. It seemed really natural to me that this was a really good way of spending a life. I was around twelve, and I was already very passionate about making pots when I met him, and it seemed very natural when he was talking about how it takes a lifetime. That seemed to me a totally natural thing to hear and to internalize. That idea of a necessary amount of attention – or attention being the ground on which you learn, seems to me perfectly natural.'

I asked Edmund if that early somatic practice of attention at the potter's wheel has informed his attention in his other fields of work, his writing and his creation of artworks. He replied that it had and that they were all profoundly interlinked. In fact he was demonstrating this as we spoke, throwing a succession of different small porcelain vessels. 'Writing and making are profoundly related to me. It seems very straightforward to me that words need the same amount of attention as objects; that words are, in some sense, objects and need to be handled carefully. The person who I read passionately growing up, who was profoundly influential to me about attention, was the poet, artist and writer David Jones who wrote much about the "thingness" of words; that each word is an incarnation in some way, a fully separate and discrete thing to be taken with enormous care. Jones not only was a poet and painter but also made inscriptions, made art out of words, so that seemed to me a profoundly beautiful idea, that words were capable of all those different kinds of attention, that they were physical in some way.'

De Waal has made an installation, a wall-mounted vitrine containing sixteen white vessels, for an exhibition devoted to David Jones, and it is titled 'if we attend', which is itself a quotation from Jones's long poem, 'The Anathemata'. Yet, de Waal said, there are distinctions in the manner of attention when throwing, writing or creating. 'They don't map each other as experiences, there are differences going on. First somatically, there is a kind of relaxation about throwing which is why I wanted to throw as I talk with you – and so I am paying attention to what I am doing at the same time as I am liberated from paying attention to what I am doing, as it is so embodied as an experience for me – as a series of movements. The great thing about throwing for me of course is that it is that it is a sitting still thing. It's not exactly *zazen* [seated meditation] but it's a placing of myself in the world, paying attention to this bit of the world. When I'm writing there is a different kind of paying attention going on. For me it is absolutely about making, I have a very strong visual/aural sense of the words. When I'm writing I am very deliberately creating a page or a paragraph or phrase, there's that kind of special awareness, attentiveness going on. But I hate to sit still when I'm writing; it's very difficult to sit still when I'm writing. So there are different kinds of tension going on. Ideally, like a kind of medieval monk, I would like to write standing up but I'm using a word processor! But there is a tension there. They are nearly aligned but not totally as experiences.'

Elsewhere de Waal has spoken about how he likes to talk, as he did with me, while making, as it takes his conscious mind out of the equation of the making. His attention, he says is in the touch: 'thinking is through the hands as well as the head'.[36] Echoing the thoughts of Matthew Crawford and Stephen Batchelor, he has expressed a concern about the lack of attention paid currently to tactility, a worry that the seemingly ingrained love for the clarity and cleanliness of abstraction, aided by the unstoppable rise of digital technology, makes it ever easier for us to live a somewhat sterile existence in touchlessness.[37]

He spoke to me of his moving out from being a potter, creating individual objects, to filling galleries with installations as being like a movement from words to sentences. 'The pleasure of moving from words to sentences takes you to poetry. It allows you to find ways of saying things, which are more cadenced, which are more ambiguous, which are not such straightforward narratives as the mug in the kitchen, the stack of bowls in the cupboard. All of which I totally love, but you can make poetry, you can absolutely make poetry with objects. That for many people that is just totally mystifying – they see lots of pots in a case and go "that's lots of pots in a case" – doesn't worry me in the slightest. It's a language and all new languages take a while, they take attention and come out of attention. Attention will happen.'

I had read earlier that de Waal considered himself 'a poet who uses pots'.[38] Music and poetry deeply influence his work. An exhibition was inspired by the poetry of Paul Celan and, as we talk, there is a beautiful late Bach cantata filling the studio. He explains: 'I am seeing some great German church spatially and hearing how it is working in that church and my hands so want to make this cantata. There is a total desire to make music with objects.' He turns to the small vessels he has been throwing as we have been speaking. 'These are going to be black pots in a long black case, with tiny fragments of gold somewhere on them. And that's as close as you can get to the *Easter Cantata* feeling of hidden gold and shadows. So there are alliances going on between sound, hands and heart all the time.'

He described to me the process of the morning, waking up to the first day of term for his children after long summer holidays, listening to a Bach cantata on the radio, playing it loudly at breakfast, thinking about our scheduled conversation, 'and this is the perfect waking piece I had to make'.

It seemed to me an ideal example of the other side of emptiness and space, the fullness and interdependence of things. Edmund agreed, but thought this was very difficult to articulate. 'The conventional models aren't sensitive enough to allow for that awareness of breathing in/breathing out, interdependence and emptiness, that filling up and emptying out, return, return, return. The feeling that is absolutely what all of this is about.' Referring to the difficulty of articulation, he said, in a way that reminded me of my conversation with James Turrell: 'That's all that you can do. You can make what you make and put it into the world. You can't force people to pay attention. So however experiential your artwork is; however much you stretch the experience in order to, in some way, make theatrical the knowledge that someone has to pay attention to something – a very long film – a place that is very far away, a building, an environment, whatever – actually people can walk away, dip in, dip out, close their eyes.'

While we were talking I had noticed an inscription pencilled on the low wall behind his wheel saying 'The Ten Thousand Things', a phrase that is often used in Taoist and Zen texts as shorthand for the arising of the multiplicity of things, for that suchness and fullness that is the other side of emptiness. Edmund explained that it referred to the title of an upcoming exhibition inspired by the composer John Cage. 'For me there is so much there in terms of his practice, of his openness to experience and – what I absolutely love is his ability to navigate the unedited and the edited within what he does. So to pay attention to everything is impossible, but if you can find a structure, if you can find the right kind of container, the right kind of structure, the right kind of formal element, then you have a way of paying attention, then you have a mechanism, a coherent way of allowing the random to become part of what you are aware of – 4'33", obviously, but in lots of his later works.[39] That idea of how do you allow the random accidental, the openness of the world to inflect into what you are making and not be overwhelmed, not to float away into inchoateness. How you navigate? That is fascinating; as a practising artist or writer, whatever you do, how do you keep growing, keep aware, keep open to things without totally losing this – groundedness in a skill, a discipline? It is a constant question.'

And so our talk came back to thoughts of time. 'A lot of it comes down to time, doesn't it, which is always contentious. How do you talk about what you do without it sounding like you are offering a model of learning or a model of your practice, without it sounding moralistic or pious? How can you talk about the fact that it actually can take forty years to learn how to do something, without it sounding terribly old fashioned and like you are straitjacketing all kinds of creativity? But there is a kind of total truth that some things you need a lot of time to do, and that the more time you take with them, the more extraordinary the experience becomes.'

Time, slowness and attention. Edmund described how in some of his exhibitions he had taken away all ancillary catalogue material and captions 'in the aspiration that people would pay more attention'. He had found this both interesting and successful in slowing down the looking, and had heard from gallery attendants that they had noticed people spent far longer walking round and round, 'not in perplexity, I hope – but actually genuinely in a slightly different place because they were being given an opportunity to pay attention, rather than fix objects and digest them and move away so quickly. The ways of slowing down the world are quite complex aren't they?'

AS THIS CONCERN with time and slowing down was shared by all the artists I had spoken with, I felt I should talk with a representative from a younger generation of artists, and one whose work encompasses both the artistic object and performance, both of which are steeped in ideas of history and the overlay of different times. I talked to Pablo Bronstein, an artist who was born in Buenos Aires but trained and lived in London. I was particularly interested in his practice, which involves complex drawings, installation and performance, and is centred on architecture and its ability to influence identity, behaviours and customs.[40] When I asked him about the place of attention in his work, he immediately saw this in terms both of his creation and his audience's reception: 'It depends whether by attention you mean my awareness of it in others when looking at my work, or my attention span in making the work. When I am making a certain sort of work I do think a lot about how much it is going to take someone to look at it and how that other person will experience the work through time, whether it's a drawing with lots of different elements or few elements or moments that grab attention and moments where the eye glides. For example, when I make a performance with ballet dancers I am aware that certain things are boring to look at and certain things are exciting to look, but of course, if you put in too many exciting things, the work becomes boring. If, for example, you have a very traditionally boring work with not much in it, or not much that is extremely seductive in it, then the audience expectations radically shift and they become a lot more tolerant. They look for certain things and they give it a certain amount of space. With my drawing, for example, the drawings that have endless detail – detail upon detail upon detail – have the same sort of readability that very simple drawings have, which is that you get their gist extremely quickly. Someone sees the work and they see a more uniform surface, whereas the drawings that people seem to return to again and again have a good balance of things that call the viewer in and things that let them not stress about.'

He spoke about the timing of works, of 'giving too much to the audience or feeling that there perhaps is too much pleasure'. He likened this to the distinction between stand-up comedians who tell two hundred jokes in an hour and those who build up and tell about three jokes in an hour – 'I think you can aim at greater depth with slightly more build-up.' He believed that the process of creating artworks and performances was not dissimilar 'in terms of their aesthetic construction. In terms of how they read over time, they are, of course, radically different. I think it's easy for someone who is trained to read a picture, even a very complicated one, in a matter of seconds, while a performance has to take the duration that it takes and you are not giving the audience the choice of knowing what it is going to look like at the end from what it looks like at the beginning.'

I found that while he was centrally involved and well considered within the contemporary art world, in other ways he was quite divorced from aspects of it. He is particularly interested in the Baroque period, which he describes as being focused from a human viewpoint as opposed to classical forms, which strive to be sufficient in themselves. The resultant sculptures or drawing that he creates refer both to past and present. They have, he says, no claim to authenticity, come from his imagination and perhaps point either towards a different and hoped-for way of being or demonstrate more clearly some of the hidden influences that were inadmissible at the time.

He is much aware of context and believes that the drawings and the performances are 'very different things to experience and to look at. I would say that the experience of looking at one of my drawings within the context of a contemporary art environment, whether it's a gallery, a museum or a collector's home or even a magazine, is slightly confrontational because the type of work I do has a lot more historical content, or a lot more handwork that goes into it than the majority of things to the left or the right of it. Essentially they are very laborious drawings, very, very detailed. Very often they are about very specific things; they may be about a particular sort of eighteenth-century architecture of a design, so that's contextually challenging for a contemporary audience. The performances fit easier within a tradition of contemporary art performances because that world is a very free-flowing world, a lot less rigid and a lot less fashion-oriented. The world of contemporary visual images, paintings, sculptures, drawings, these things are governed so much by fashion that anything that feels a little bit different from this feels like a radical break. Whereas in the performance art world the parameters and the context are a lot more allowing.'

When I asked him if he felt that context makes a difference to our attention, he replied: 'I think it really does. There is also something that I think is being lost from a lot of contemporary artmaking. That is, it's a sort of competent illusionism – the ability to make you look at something and in that space, in that area of canvas say, see further space, or see greater depth or see another world. That sort of thing has been replaced by an immediacy of objects and experience. There are people making very complicated and conceptual works, but they are, what I find, hieroglyphic. You can read them from left to right, or perhaps right to left or up and down, but it's like accumulating a series of visual signs. The tradition of Western art from the Renaissance onwards was to create a stage set where you could place yourself in, and that, I think, has radically changed.'

He cited Dan Flavin as an exception and one who absolutely 'creates stage sets and creates an unbelievable sense of space and illusion', but he felt that 'a lot of art at the moment is very literary based rather than optically based. You find that a lot when you visit an art installation, you get an overall message but it's through encountering various elements that then lead to another meaning and that feels very literary to me.'

Touching on another aspect of time and attention, he spoke of his appreciation of art that has been created with a lot of care. 'It's when humans take a long time over an object and they labour over it, and they put a lot of thought and hours making something with their hands in a very individual way, we as other humans tend to look after it better and for longer. Over time you see this is the case. For example, we have more rare, glamorous, extremely beautiful objects in museums than we have, say, shoes from the period, or wooden spoons from ancient Egypt. So the more mass-produced and the more quick an object is the less we tend to value it after it's made. I think that there is a question of attention in that what a lot of these objects have in common is a highly worked surface, and I get inspiration from them. Very often they have little scenes or vines or ornaments, and in these places I somehow meditate, if that makes sense?'

Pablo spoke also about space, describing how he and his partner had moved to the seaside of east Kent a few years back 'in order to expand visual horizons. We were finding that there is a sort of claustrophobia in London that you get very used to, and I was noticing that my spatial awareness was starting to go. To give you an example: London gives you a vista in which the furthest away you can look, apart from the top of Hampstead Heath or on the River Thames, is about a block and a half away, if that, and in certain areas all you can do is look across the street. When we moved to the sea, I think we moved because there is something extremely rewarding about a horizon line, about seeing very far. It took me a few weeks to accustom myself to that end line, because my eyesight just wasn't registering it somehow. I find that a lot of the detailed work that I do, every so often needs a very blank counterpoint where I sort of stare out at something that feels as if it goes on for longer than the page that is half a foot away from me.'

ALL OF THESE ARTISTS are concerned, though in quite distinct ways, with some form of enhancing our experience and defamiliarizing the habitual. Though the language they use to describe it is different, it all involves attention in all senses of the word, attending and tending – in the act of creation of the artist and in the act of reception on the part of the audience. By paying attention to the world, to their process of perception, to their own habits and those often-unthinking and unthought habits of the culture around them, and to their experience of time, artists help to train our attention in new and creative fashion. Some months after talking with both Garry Fabian Miller and Edmund de Waal, I listened to them talking to one another at a public lecture given during Photo London 2016. Far from his home on Dartmoor, Garry emphasized, as he had done talking with me, the centrality of walking for his work, describing it as 'the central practice of my life'. He spoke of it as 'walking into a thinking space, and sometimes something happens and sometimes it doesn't – the possibility in the walking and the things I am going to see are there. If I don't do it, it's not going to happen. So the repetition is that every day the possibility is there of finding what you are looking for, so it's an anchor state and if you don't put it into practice it isn't going to happen. It's a kind of notion of the Protestant work ethic, but there's also a mystical belief in discovery.'

At the end of the discussion de Waal summed up that mixture of hard repetitive practice and the hope of epiphany as 'Protestant, and mystical, hard work, and the integration of vision and keeping going in the sense of a complete life.'

8

Aural Attention:
Listening and Hearing

For me, it all begins with listening.

Where the eye divides, the ear connects. Auditory space is a
denser, more fluid medium than time and space as we usually
experience them. In audible space, everything is connected,
and the listener is always at the centre.

JOHN LUTHER ADAMS[1]

Music has been described as the most pure artistic form, being
symbolic of nothing other than itself, a fact which George
Steiner has described as the 'radical untranslatability of music'.
He describes it as escaping the limitations of language, which by
its very bondage to linearity, causality and sequence in time, necessarily abdicates from 'the manifold and the self-contrarities of
the world . . . handcuffed to the avarice of logic, with its ordinances
of causality, with its (probably crass) segmentation of time and
perception into past, present and future, identity'.[2]

Freed from such linguistic limitation, music and dance are 'primordial motions and figurations of the human spirit which declare
an order of being nearer than is language to the unknown of creation'.[3] Aldous Huxley, who in his novel *Island* had the birds call
us to 'Attention, attention', would seemingly agree. He wrote many
years ago that

From pure sensation to the intuition of beauty, from pleasure
and pain to love and the mystical ecstasy and death – all
the things that are fundamental, all the things that, to the
human spirit, are most profoundly significant, can only be

experienced not expressed. The rest is always and everywhere silence.

After silence that which comes nearest to expressing the inexpressible is music. (And, significantly, silence is an integral part of all good music.)[4]

From a different perspective, composer John Luther Adams would certainly seem to agree that music is primary and closely linked to attention:

Quantum physics has recently confirmed what shamans and mystics, poets and musicians have long known; the universe is more like music than like matter. It may well be that our most fundamental relationship to the great mysteries is one of listening. Through sustained, concentrated attention to the fullness of the present moment, we listen for the breath of being, the voice of God.[5]

Philip Glass touches on another feature of art, most pertinent to music, which is one that others with whom I have spoken have mentioned. It is also a significant feature of the way we see the world when we pay close attention – its interdependence: 'The important point is that a work of art has no independent existence. It has a conventional identity and a conventional reality and it comes into being through an interdependence of other events with people.'[6]

In performance, or at least public performance, music clearly exhibits that interdependence. Even more than the collaboration between writer and reader, music is dependent not only on its audience, the listener, but also on performer(s) and conductor. As Alfred Brendel has beautifully described it from the performer's perspective: 'In the concert hall, each motionless listener is part of the performance. The concentration of the player charges the electric tension in the auditorium and returns to him magnified; thus the audience makes its contribution.'[7]

Both Adams and Glass have been profoundly inspired by John Cage, whose influence in every field of art practice has been enormous. Composer, writer and artist, Cage has been mentioned many times already in this book. His work and his life exhibit his constant concern with attention, emptiness and contingency, and his mission to evade habit and to open our eyes and ears free of expectation, like and dislike. He is perhaps best known for his still shocking work 4'33", in which a pianist comes to the stage, sits down at the piano, opens the lid and remains silent and static for the given duration of the piece, merely opening and shutting the piano lid to distinguish the score's three 'movements'. It was, and is each time it is 'played', a demonstration that there is no such thing as absolute silence, that silence is just the background to sound, and an invitation to attend without discrimination. (That such a demonstration itself can be misunderstood as time passes is perhaps shown by a 2007 case in which a performer of this work was threatened with legal action by the guardians of the Cage estate for performing it without permission.[8]) Cage himself said of 4'33"

it opens you up to any possibility only when nothing is taken as the basis. But most people don't understand this as far as I can tell . . . But the important thing . . . is that it leads out of the world of art into the whole of life.[9]

In other attempts to avoid habitual response, he wrote music whose progression was decided by the throwing of coins according to the ancient Chinese I Ching, prepared pianos to produce unexpected sounds, and arranged text unusually on the page. There is not a corner of the contemporary art world where his influence has not been felt and his shadow fallen. As Glass explains:

What Cage was saying is that there is no such thing as an independent existence. The music exists between you – the listener – and the object that you're listening to. The transaction of it coming into being happens through the effort you make in the presence of that work. The cognitive activity is the content of the work. This is the root of post-modernism, really, and John was wonderful at not only articulating it, but demonstrating it in his work and his life.[10]

Adams sees it the same way:

> One of the most far-reaching assertions of Cage's work was the notion (still considered 'radical' in some quarters) that music depends on listening. When we listen, the whole world becomes music. After Cage, Western music would never be the same. The center of music is no longer the omniscient composer. It's the listener. And the composer is now free to be a listener too. The broader implications of this musical worldview are ecological. Cage taught us that music is Nature and Nature is music.[11]

Thus Adams today writes, 'It's only through the presence, awareness, and creative engagement of the listener that the music is complete.'[12] His own desire for greater freedom for discovery by the listener, the musicians and the composer has led him into new musical territory – to composing music that from its very inception was intended to be heard outdoors. 'Making music outdoors invites a different mode of awareness. You might call it ecological listening.' He believes that rather than being invited to focus our attention inward as in the concert hall, we are challenged to expand awareness to encompass a multiplicity of sound, to listen '*outward*. We're invited to receive messages not only from the composer and the performers, but also from the larger world around us.'[13] Perhaps the most radical expression of this is a 'piece' entitled *The Place Where You Go to Listen,* which is installed in a room at the Museum of North Alaska at the University of Fairbanks. *The Place* translates the actual real time occurrences of earth and air into music: data from seismological, meteorological and geomagnetic stations in different parts of Alaska are fed into a computer and transformed into sound. Adams has said of this:

> This is a space to hear the unheard music of the world around us. The rhythms of sunlight and darkness, the phases of the moon, the seismic vibrations of the earth and the fluctuations of the earth's magnetic field all resonate within this space.[14]

It is perhaps the closest one can come to a suggestion that Jane Hirshfield made to me of listening to the attention of a plant. When I spoke with Jane about Adams's aural installation, she responded to it as 'the world's collaboration'. Another piece of his music, *Inuksuit*, in which an outdoor audience walks freely through and around the musicians 'as waves of sound pass from one side to another, contemplating the collision of music and natural sound', has been described as 'pretty close to pure musical mindfulness'.[15]

Following the Cage trail again, Adams says the title of his 2015 Pulitzer Prize-winning piece *Become Ocean* was

> stolen from John Cage, from a little mesostic poem that he wrote in honor of Lou Harrison in which he compares Lou's music to a river in delta, with all these different influences and currents, coming together in a big beautiful sweep of music. And in the last line of the poem, Cage writes, 'Listening to it, we become ocean.'[16]

Far from the wild and space of Alaska with which he has been so associated, I was able to discuss all these topics in a meeting with John Luther Adams at his relatively recent home in New York City. I had opened our conversation by explaining that earlier, writing about emptiness, I had felt that people who embraced the concept from outside Eastern traditions seemed to come to it through attention. I said I felt that we all think we know what attention is, but we actually don't pay much attention to attention itself, only to its content. John responded: 'This for me is where emptiness and attention come together. I'm much, much less interested in content than I am in awareness.'

When I asked, 'What does attention then mean to you?', he replied: 'Well as I often say, music is not what I do, it's how I understand the world, so when I think about paying attention it is almost always through my ears. It all begins with listening. In some ways I'm absolutely scatter-brained but when I'm listening or when I'm in the act of trying to receive music, when I'm composing, I'm very focused. I'm paying attention in a very deep way that I wish I could embody in the rest of my life. But for me music is a call to attention. It's an invitation to slow down, yes to focus inward, but also at the same time to listen outward and just to be fully present in the moment and in the place, and as you may know the place is a central obsession of my life and my life's work. This is not just in a metaphorical, poetic sense, but I'm talking in a very physical geographic, ecological, experiential sense. I want to know where I am. I think one of the most powerful ways we can know where we are is through listening and if we understand more fully, more deeply and more broadly where we are then I think we understand, or we can begin to understand, more fully, more broadly, more deeply, who we are, and where we fit into this miraculous world that is the only home we know.'

When I suggested that in that statement there seemed to be both a movement of getting to know self better and one going beyond self, John agreed: 'Yes, for me music is a way to transcend myself, or go beyond myself and reconnect with something bigger and deeper and older, and in some way unfathomable. I grew up all over the United States and as a result never really knew where home was. People would talk about going home and I never really knew what that meant. But when I was 22 in 1975 I first set foot on the ground in Alaska and immediately knew that I had come home, that I had found a place to which I could belong, deeply. I lived up there for most of my life and Alaska allowed me, encouraged me, to imagine that I could somehow draw my music directly from the earth; that as an artist I could work somehow outside the culture and more directly in touch with the sources of all human culture. Of course that's a patently absurd idea, we are inescapably products of our cultures, but it turned out to be a very useful conceit which I still follow as best as I can in my life's work. So when you were talking about your first book and this connection between emptiness and attention, my understanding of that is not through any kind of ostensibly spiritual practice – I don't meditate. I flirted with it early on. I window shopped in various religious traditions and never found anything that felt like home, never found anything that I could believe in that deepest sense. Obviously I found beautiful resonant elements from cultures and traditions all over the world, and perhaps I felt the deepest resonance with what I understand of Native American spirituality and lifeways and traditions, but again I'm someone else, a product of another fragmented culture. So in Alaska my experience of emptiness was not through a religious practice, it was through my experiences in wild country, in original landscapes.

'When you are out in what I sometimes call the real world – you damn well better pay attention or the glacier will calve and sweep you away or the bear will spring out of the alders and swipe you or the storm will descend and blow you away. So there's a very direct and physical, very human and very ancient experience that we have of feeling ourselves as part of what we call nature and yes, we are inextricably part of nature, but you can be out there in the Arctic, on the coastal plain or in the mountains and really the mountains don't care if you are there or not, and they don't care if you live or die. Sometimes I found myself on the razor's edge between beauty and terror that Edmund Burke called the sublime. And for me that's about as close to religious experience as I get and that's right at the heart of my life's work and my practice as an artist.' As John was describing the effect of the landscape of

Alaska, ironically we were brought back by the howling of sirens to the immediate presence of the streets of West Harlem.

I spoke of my three themes of attention as a skill and one that can be practised and enhanced in many ways, whether by walking, meditation or music, and John agreed with this. In response to my third theme of some sublime, transcendent experience attained through attention, he suggested that in a world where the norm is a deeply fragmented attention, 'there is not only a spiritual imperative but also a biological imperative to pay attention . . . in a profoundly biological sense it's an imperative of survival, not just our survival as individuals, but now, at this perilous moment in our history as species, attention is essential to our collective survival. If we don't wake up and pay attention to the world around us, we are not likely to be here much longer. Talk about survival; how many people get hit in the street in New York City because they are texting. That's symbolic I think of our whole predicament at the moment, as human animals particularly in technological society. So this is a matter of survival.'

I said that for me that aspect of getting beyond the blinker of the self, and understanding, as he had so beautifully described it, that the mountains couldn't care less about us, was profoundly important, and he replied that this for him was profoundly reassuring. 'That is like being in the presence of Yahweh or whatever you might call it. I want to be lost. I want to experience fully my fleeting nature, my fragile existence. For me that's right at the heart of what you might call religious experience. That's much more comforting to me than somehow feeling I'm in control of everything, including myself. You used the word transcendent and one of my earliest heroes, role models – perhaps if I could blame one person for the wrong direction of my whole life it would be Henry David Thoreau. I encountered Thoreau very early and was deeply impressed by his misanthropic idealism and I went to Alaska and lived out my own Thoreau fantasy in my own private Walden in the woods there. So I guess in some way I am part of the American Transcendentalist lineage that musically began with Charles Ives. I always wondered, and I have had this conversation many times with the composer Kyle Gann, who has just written a whole book about the Ives Concord sonata. I had never been to Walden and a couple of years ago I was teaching at Harvard and we decided to go to Walden together, and we walked around the pond together and we visited Emerson's beautiful gravestone and Thoreau's pitiful little rock, and we talked all day about those guys in particular. Kyle sees himself as an Emersonian, and I am very much a Thoreauvian, but my question for that day was "What are we supposed to be transcending? If we are Transcendentalists, what are we transcending?" We need to come down to earth; we need to be more deeply embedded into it. Something you said earlier made a little light go on for me, and that is maybe what we are transcending is ourselves – and I'm OK with that.'

I proposed that perhaps what is most important is the way we look at things rather than what we look at, which brought us back to the quality of attention. I quoted Iain McGilchrist's suggestion that the soul might be a quality of attention, which to me loosens its religious and otherworldly, idealistic trappings. Adams thought this was a beautiful notion, and continued to talk about a big piece that he is currently contemplating, 'a sort of lamentation and celebration, ultimately for all of us and all disappearing things in this world'. He thinks it may be focused on birds, and thinking of extinct birds and endangered and threatened birds, and ultimately all the birds, he had thought he might be talking of a requiem. So he had gone to the Latin text of the requiem but had found that there was very little in it that he felt he could grab hold of and say 'I could say this', 'I could sing this' meaningfully today. 'How could I possibly transform Christian mythology into something that feels like mine?' He felt that times had changed considerably since Benjamin Britten had written the *War Requiem*.

From this our conversation returned to emptiness, as I asked him about a quotation that Kyle Gann had written about him as 'latching onto white as a pervasive metaphor'. He had also written himself in *Winter Music* that white is not the absence of colour, it is the fullness of light, and silence is not the absence of sound, but the presence of stillness, which so clearly expressed what I had been trying to describe in my own book on emptiness; that it is fullness, the possibility of everything. Laughing, he broke in: 'It's complete. It doesn't need us to place anything within it, especially if we are paying attention. Yes, it's interesting – throughout my creed of life there has been this process of emptying out. You can look at the progression from *Clouds of Forgetting* to *In the White Silence*, to *For Lou Harrison* to *Become Ocean*, and the number of topologies, or the number of musical elements, the types of musical material in those pieces goes from 4 to 3 to 2 to 1. Because I want to be in it, I don't want to be told about it. I don't want to watch a movie. I don't want to read a novel about it. I want to be in it. I'm happy to be told a story, music can do that beautifully, but it's not what music is about for me. I want to live in the music. I want to live through the music. I want to inhabit the music.'

In *Winter Music* he had described how one way of losing perspective in order to gain inhabitation was to lose melody and harmony, which constituted a kind of musical perspective. Interestingly, in the book, he had compared this to some of James Turrell's immersive *Ganzfeld* works. Now, he said, 'I would add to that a more multi-dimensional approach to musical time, not measuring out tick, tick, tick – one, two, three, four – unmetered time. Morton Feldman said, "You want to experience time not as the animal in the zoo but as the wild beast in its native habitat", or like Samuel Beckett talking about "time undisturbed". I want that sense of the unending present and presence of time, but also the multi-dimensionality of time in the same way that we experience near and far and middle ground and behind, the same way that we experience physical space. I want to experience time flowing at different rates simultaneously. So originally all of my music encompasses multiple streams, multiple layers, multiple rates of musical time, and then

there is space itself, which is becoming more and more a fundamental compositional element for me, not just a metaphor. Well,
I'm just working on a big new orchestra piece called *Become Desert*,
which completes a trilogy – *Become Ocean, Become River* and
Become Desert. And oftentimes I begin with a new musical nota-
tion or maybe a drawing of a form or the shape of the music as I
imagine it, or maybe a few words or a title or poetic imagery that
helps me to begin to discover the music. But for *Become Desert*,
the thing that came first, the most fundamental element of this
new piece, was not just the instrumentation, not just the numbers
and the types of the instruments of the orchestra, but the actual
physical employment of the instruments in the space. It's five
different ensembles; placed around, above and behind the listener.
And once I had mapped out where the sounds would come from,
the music started to flow. So space itself has become a fundamental
compositional element for me, in the way that it might be for a
landscape designer or an architect or a sculptor.'

I suggested that perhaps the space he originally started with in
his music was a response to the physical space of Alaska and was
a kind of outer space was now so embedded and embodied that
now it was an inner space that could be expressed in the music
going out. To which he replied: 'Yes, and I think that's why I was
finally able to leave Alaska – I never thought I would leave Alaska
– and maybe why I *had* to leave Alaska. I remember, several years
ago now, I began working outdoors. It finally occurred to me that
after forty years of making music inspired by the big world but
intended to be heard indoors in the small world, it was finally time
to step outside and make music that was intended from the start
to be performed, heard, experienced out of doors, and that
changed everything and it drew me to understand the physical
space, the musical space, the acoustical space, the performance
space, you might say the ceremonial space in new ways and led
me further away from space and place as metaphor towards space
and place as literal reality, lived, experienced, sounding space,

experienced geography, and that's led me I think to let go of the more specific sources of my work in Alaska, and I think that's because, as you suggest, Alaska now is just so deeply a part of who I am and what I do that I carry it with me wherever I go and whatever I do. It is not site-specific. It is what Robert Irwin might call "site-determined". This is no longer music from or about Alaska, this is music from, about and of, here and now, wherever that may be and this allows me now, I have to say, to feel at home in New York City. I love being a beginner. I had said when I was young that I would never leave Alaska and imagined that I would die there, and maybe I will before it's all over, but I always said that if I ever left Alaska it would be for the desert, and I imagined it would be New Mexico or Southern Utah. I never imagined it would be the Sonoran Desert of Mexico. But it is, and it's wonderful, and there is in that sense of apparent emptiness and space, there's a real resonance, a similarity with the tundra. It makes sense: I felt instantly that I knew where I was down there in a way that I might not in, say, a rainforest. But at the same time this exhilarating feeling of being a beginner, of being a child, of starting over, of learning the light and the rainfall and the clouds and the birds and the plants and the rhythms of night and day and where the stars are. The sky is very different. And I love that. That sense of discovery, of learning. It's really where I want to be. We have lived down there now for eight winters, and I have done some of my best work there in the desert in Mexico; a piece called *The Canticles of the Holy Wind, Beyond Ocean* was written there and *Become River*. I have also just finished the first complete draft of my memoir down here.'

I said that I felt both desert and tundra could be described as spaces of great potential and emptiness, even as white spaces, via discussion of Edmund de Waal's book on porcelain and white, and we returned to the idea of white as a metaphor. 'A metaphor,' John said, 'for that place where emptiness and fullness meet.' And when I referred to Turrell's *Ganzfeld*, or sky-watching spaces or even

Cage's 4'33", he said: 'Of course. Let's go there, it may be relevant to your work. We will get musical without getting too technical but I think it may be pertinent. Turrell is one of my favourite artists and I feel a great affinity with his work. There is one piece of mine called *The Light Within* that is inspired by his sky space over on PS1 here in Queens and I have been touched that he has asked that that piece has been performed at openings of several of his sky spaces around the world.[17] We have never met but that makes me feel that he feels some reciprocal affinity, which is lovely. Talking about white as both fullness and emptiness and talking about 4'33" suggests two sides of music, two fundamental elements of music, certainly two deep sources for my music – noise and silence – and really they are the same thing. Cage said, sometime in the '30s, most of what we hear around us most of the time is noise, it's broadband noise not musical tone – like the roar of the city if we open the doors now, you might hear the ice cream truck go by or you might hear someone playing music too loud out in a car, but most of what we hear outside we would not call tone, we call it noise, and the same thing is true in the forest. You'll hear a bird singing, or the wind or the roar of a waterfall, but most of what we hear when we are in the world around us is noise and, as Cage said, when we regard it as an interruption, it annoys us, but when we choose to listen to it we find it endlessly fascinating. So that's one lesson from Cage that we have learned.

'And the flip side of that of course is his invitation to pay attention to what we regard as silence, as shown in 4'33". And as Kyle Gann says in the title of his book about that piece, *There is No Such Thing as Silence*. When you are listening carefully there is always something to hear. So my music encompasses – there is a piece called *In the White Silence*, which is 75 minutes of continuous music and not a moment of silence in it. There's another 75-minute piece called *Strange and Sacred Noise*, which contains within it several minutes of composed silence. So what's up with that? Did I get my titles switched up? I don't think so. I think, without understanding it fully, I must have realized that noise and silence really are parts of the same whole and they lead us to the same conclusion, which is when we are listening, when we are paying attention, the whole world is music.'

IT WOULD SEEM that this awareness of sound and silence, attention, intention and listening, is a feature of contemporary music. In a radio programme considering the work of composer Morton Feldman, someone whose work was much influenced by his friendships with many of the notable twentieth-century painters in New York, it was said that 'every note is determined, every mark and notion is designed, to focus the performer's attention and us as their listeners'.[18] Feldman's quiet music focuses the attention of the listener and, in the words of pianist John Tilbury, 'rather than expressing a form, his music creates a space, a kind of release for both performer and listener providing an antidote to the congestion that blights our lives'. Feldman himself had said:

> Sound for me is the experience. I'm looking for the experience in what happens with the sound, which is telling me how it is formulated, rather than looking for the experience in the system. I am more immediately involved with the sound. I am in the sound, so to speak.

He also spoke of the importance of the sound dying away, echoed in the life and death relationship: 'It's the decay that is the true life of the sound.' Composer Kevin Volans believed that Feldman was involved with the process of making music rather than the finished piece of music.

In an extraordinary collaboration of visual and aural attentiveness, Morton Feldman wrote a piece of music for the opening of the Rothko Chapel at the University of Houston. Mark Rothko had designed a series of murals for the interdenominational chapel that his son later described thus:

The chapel murals exist at the intersection of everything and nothing. They are silence and a full-voiced shout. They are concrete and yet hardly there. They speak simultaneously of the never was and the ever shall be. In short, they live in the realm of the paradoxical, or the mutually exclusive.[19]

Writing of the musical work, Christopher Rothko notes that Feldman, like his father, was both architect and artist; that just as his father understood the way the space around his paintings shaped them, so Feldman showed that it is the space, the silence around the notes, that shapes the form of the music. 'It is the silence that gives the notes their meaning', so that 'Feldman teaches us that silence *is* sound, just as Rothko shows us that emptiness has substance.'[20] Both call our attention to the absence of other things, of noise and of narrative, so that we can appreciate 'just how rich that absence is, and just how much voice, how much music, there is within us'.[21] He also writes: 'Feldman's Silence = Rothko's Space.'[22]

This reminds me of a programme note that I had read at a performance of Feldman's Rothko Chapel piece by the San Francisco Symphony Orchestra in 2011 in which the writer had described Feldman as

the ultimate low-talker among composers, and he can make you strain to listen. When you have to work to hear somebody, you tend to pay undivided attention to what he or she is saying. So it is with Feldman's music: it demands your concentration.

He went on to describe the chapel itself as a contemplative space, which Feldman's music approaches in a similar spirit of attentive mindfulness.[23]

COMPOSER AND TEACHER Pauline Oliveros has been instrumental in bringing attention to the process of attentive listening, creating a practice called Deep Listening. This began with a composition called *Sonic Meditations*, which she has described as a body of work that could be performed by those without any musical training, as pieces that are 'ways of listening and responding'. She created these practices as a response to the observation that many musicians were listening inadequately to their performance in terms of global listening 'to the space/time continuum'. This disconnection from the environment, which included the audience, prompted her to investigate attentional processes. Through such investigations, she learned how she could affect her emotional state by concentration on a tone; how she could affect her responses of relaxation or tension; and how she could attain a heightened state of awareness that gave her a sense of well-being. Such investigations gave birth to the practices of Deep Listening: 'to hear with thoughtful attention'; 'to give attention to what is perceived both acoustically and psychologically'. Oliveros describes such intense listening as a form of meditation, a practice in which attention is directed to the interplay of both sound and silence, to the whole space/time continuum, which results in an expansion of consciousness:

> **Compassion** (spiritual development) and **understanding** comes from listening impartially to the whole space/time continuum of sound, not just what one is presently concerned about. In this way discovery and exploration can take place. New fields of thought can be opened and the individual may be expanded and find opportunity to connect in new ways to communities of interest. Practice enhances openness.[24]

A SIMILAR EMPHASIS on listening and attention was discussed when I talked with two musicians, husband and wife Sam Richards and Lona Kozik, performers and teachers of improvisation.[25] Sam, a pianist, teacher and John Cage biographer, said of improvisation, 'For the most part the training is training in attitude of mind.' He described teaching improvisation by instructing his students at the beginning just to limit the material to a drone and not move off it until it moved off itself: 'A friend of mine once said to me, "You don't play what you want to play, you play what the music wants." But if you do play what the music wants you will find out that really is what you wanted anyway. There are these states of mind that you can get into whereby you are concentrating so heavily on the thing that the you part of it can take a back seat. It doesn't mean to say that it's not happening, that it's not you, but I do think that there is that in it.'

Lona related that to her own experience, saying that she was constantly trying to get people to move from worrying about technique and playing to considering the technique of listening. 'That's the difference. You say, "Let's take the playing problem out of the equation and everybody can drone on G." It's easy, so they stop thinking about their technique and everything they have learned about their technique, and that direction of saying "hold the drone until it moves off" also takes the intention away. It's almost like this inner intention – "I'm going to play all these things I know how to do – I've got to show them I can play." Everyone has to get past that. And so that's really the technique for saying "we are only going to have a technique of listening". In improvisation it's possible you will never play but you are improvising brilliantly. It's flipping the intention.'

She described an exercise where the players are asked to match one another so closely that you can't hear anybody's individual entrance, 'then you are really engaged in listening'.

Sam explained that he also used a formula he had taken from Japanese Buddhism, inviting students to pay attention in relation to the three realms of self, other and environment. Firstly to listen to themselves, then to specific others in the group, 'and then the final thing, the environment, can be in two ways. One way would be the whole ensemble, and the other way with the environment is – what is the environment, the actual physical environment, the space and how might it induce you to play, and might there even be other noises going on.' This threefold configuration of attention reminded me of Maura Sills's description of the attentional training for Core Process Psychotherapy.

Sam and Lona spoke of the distinction between jazz with its mostly solo improvisation and free improvisation with its collective approach. Sam said: 'I think that in any genre of music, you only do really well in it if you are willing to go past received wisdoms and disciplines and take risks, go past your training, even free improvisation training, think that when people become very reliant on what they already know, because there might be an insecurity there – and even the best players in the world have their insecurities – it does become really stale, unless they are willing to go past that into territory that is not known.' Together Sam and Lona quoted Sun Ra: 'Don't play what you know, play what you don't know!'

'That', Lona said, 'is where attention comes in. Because you can rehearse to a point where you don't pay attention any more and sometimes that is part of what you are trying to do in a rehearsal because you think, "I need to be able to play this in my sleep," because when you are in front of people you don't know how your nervous system is going to react. You need to be reliable in one sense but you can see how really great performers (even classical musicians) are the ones who are willing to go that one step further, take the risk in front of people. You can see when someone is obviously very well practised, but there is something about the performance that doesn't really give us anything, it's flat. I just think that perhaps it relates to attention because if you have practised enough doing something and you know it's reliable, then you don't have to pay attention any more in a certain way, and it's only the ones who are willing to keep pushing – maybe they are confident they can rely on their technique and everything they have worked for – but it's almost like – any practice – free improvisation is kind of opening everything up but even that has its own conventions and practices and you have to get past those as well. . . . [This] makes me think of Evelyn Glennie, the percussionist who is deaf. She talks a lot about listening with the whole body, the idea of opening everything out. That's true of doing a conventional concert, of being able to read a room and a situation. Which might even include the weather. It's like the more conventional or institutionalized something is we don't think of that. We are all busy thinking of the basics, the tasks, but really what we should start with is learning to listen.'

Sam spoke of the resistance there always is 'to letting go to just about everything that defines you as a musician, your technique, your concepts of beautiful and ugly and the definition of taking risks. It depends entirely on context. There is a certain way that training improvisers does overlap with meditation and perhaps psychotherapy too.' All are practices of attention, journeys towards openness. Lona felt that 'as long as I was doing that kind of process regularly I could handle anything else in life. It was very stabilizing.' And Sam thought that what we had been talking about with respect to improvisation 'applies to every other form of music making, classical ensemble, not so much an orchestra for there the conductor takes responsibility for most of the decision-making, but even so you have to be listening. Or in a jazz group or folk band, there has to be a quality of really focused listening. I can't imagine a string quartet without it and it's like balance, simple things like if one player is playing too loud there is no balance. Then also, when you are playing a piece of classical music you are expected to interpret it, but if the piece becomes all about your interpretation and nothing else, it very quickly becomes boring.

'I think also with relation to attention, that there is this conventional distinction between the creator of music, the performer of music and the listener. And in improvising the first two are the same but in composition of music it is also a question of attention – it's got a lot of the same dimensions to it as improvising has to the extent that some people confuse the two. I'm not convinced of that, but I am convinced that if you are improvising it's all happening in the way you relate to what the music wants rather than what you want, and that kind of thing also happens in composition. You must pay attention to the material because it might actually be asking to go in a certain way which may not be the way your ego wants to go, but you will make stronger music if you follow where the ideas are taking you. So that's another kind of attention. It's not so much on the spot as you can go back and erase it. In my own composition I have enjoyed the exercise of wondering where does this want to go? We have talked about performing but there is also the attention of listening to music too and for us as musicians, I think a very big part of the practice is to sit and listen to the music with full attention. We have so little opportunity to do that, we put it on in the car or around the house; we have this sort of half-attention, but there should be some time for full attention. But it's a lovely exercise to put on some music and truly pay attention and it's surprising how like meditation it is – how you see the attention drift.

'One of the things of my drone exercise was that it went on for a long time and I would risk people getting agitated as I thought it was very important to deal with these very big lengths of time. But I still feel that if the improvising is really good, and by good I mean that you have paid attention throughout, that time sort of disappears and I think it takes you into a realm that perhaps you could call the sacred. Mircea Eliade talked of the sacred experience of timelessness, about the time before time in mythology. There is a sense in which you forget time. If you are praying or Buddhist chanting and you are really tuned into it, you don't know if you've been going for ten minutes or half an hour. It happens fairly rarely and most of us don't get to that exalted state, but you can with music, you can with improvisation and I'm pretty sure that if I come out at the end thinking "I really didn't know how long that was," that there has got to be something to it that took me to a sacred realm. I think it's really important, and the problem with the short attention span is that you don't get that deeper, quasi-sacred experience.'

WHAT ALL THESE COMMENTATORS have in common is an openness that moves beyond expectation. As Cage said, music is sounds, and to demonstrate this appropriated 'sounds' to be his music, not form or structure. Returning to Paul Klee's distinction of the second type of attention that is the mode of the good artist, an undifferentiated attention that Ehrenzweig calls 'the full emptiness of unconscious scanning', creative aural attention requires this polyphonic awareness of sound, this oscillation between focus and unfocus, differentiation and undifferentiation, to listen to sound undisturbed by name and association.

9
Embodied Attention

There is more reason in your body than in your best wisdom.

While intelligence was formerly concerned merely with cognitive intellect, talk of emotional intelligence is now common, yet attention to somatic or kinaesthetic intelligence is still rare. Kinaesthetic awareness refers to the sensation that informs us where our body and limbs are in space and how it feels to move. Possibly this lack of recognition reflects the lesser value given to matters (and matter) of the body since the Cartesian division of mind and body. Thankfully, however, there are those who are making it their concern to change such perception or, more pertinently, the lack of it. Some come from the fields of healing, others from the creative arts. We earlier learned that Antony Gormley bases his sculpture on the idea of the sculptural object not as representation but as reflexively affording a participatory experience, returning attention to our own occupation of space.

In the Norman Doidge book to which I referred earlier, there is a fascinating chapter on the work of Moshé Feldenkrais, whom Doidge describes as 'one of the first neuroplasticians'.[2] Feldenkrais devised a method of healing that relied on the fundamental contention of the inseparability of mind and body from which arise these core principles:[3]

1. It is the mind that programmes the functioning of the brain. While we have a number of hard-wired reflexes, the organization of our brain is experience dependent and may be rewired – in short, neuroplasticity.

234

2. A brain cannot think without motor function. Motor movement, thought, sensation and feeling are inextricably linked.

3. Awareness of movement is the key to improving movement. This focus on attention and monitoring of experience arose from Feldenkrais's experience of Eastern meditation and martial arts. It both precedes and reflects Michael Merzenich's later research findings that long-term neuroplastic change occurs most readily when learning is accompanied by close attention.

4. Differentiation builds brain maps. Brain maps are the representations of the body as held in the brain. (Fascinatingly for those who have any familiarity with a wide range of Eastern practices, neuroscientists describe these as a 'virtual body', a term used in many spiritual practices.) It reflects the fact that it has an existence in the mind and brain that is independent of the physical body. Following injury to a body part, the representation of that part in the mental map becomes smaller or even disappears. By making the smallest possible sensory distinctions between movements, and attending to them closely, patients begin to experience them subjectively as becoming larger and reset their brain maps accordingly.

5. Differentiation is easiest to make when the stimulus is smallest. With a tiny stimulus, minute difference is more easily noticed, and it is the awareness itself that helps reorganize the brain.

6. Slowness of movement is the key to awareness, and awareness is the key to learning.

7. There should be no great effort. As Doidge puts it, rather than 'no pain, no gain,' it should be 'if strain, no gain'.

8. Errors are essential. There are no 'right ways', only 'better ways'.

9. Thus random movements provide variation that may lead to developmental breakthroughs.

10. Even the smallest movement of one part of the body involves the whole body.

11. Many movement problems and their accompanying pain are caused by abnormal habit rather than abnormal structure.

Catherine McCrum was first a personal trainer in the field of fitness and sport, then became an accredited practitioner in Feldenkrais method, and is now also pursuing training as a psychotherapist in the Gestalt tradition. She talked to me about the importance of attention in her practice, saying that in the way she works attention is the most important and the most interesting thing for her, 'and it's also in some ways, the aspect of the work that people find the most challenging when they come. Mostly people come because something hurts or they want to make something better or they want to improve their performance or coordination, and they want to do it in such a way that they really don't need to pay attention, but somehow it's magically done for them – a magic wand kind of thing.'[4]

While what drew her into the work 'was the mindfulness aspect; when you're present in your body attending to how you move, and to the relationship of different parts of yourself, you are present to internal sensations', she finds that for other people this is not the case. They 'are often very attentive to their thoughts or ruminations or the arguments they have in their heads with the different aspects of themselves but bodily attention, in my experience, is not such a big aspect of how we live'.

She explained that 'Feldenkrais works on making very small subtle differentiations. As an example: a principle that he had was to reduce the effort in order to feel more. You might direct someone lying on their back to push with their foot and roll a little to one side. How you are looking to direct them is to do less in terms of muscular effort, focused trying effort. Then what seems to happen is that the less forcing they put into the movement, the more they feel, so that their sensory perception, their proprioception, improves enormously. And when you can feel more clearly how you are moving, you have more choice about how you can do it.'

Feldenkrais was developing his work in 1950s, '60s and '70s and was a contemporary of Fritz Perls, the originator of Gestalt therapy. It was also a time when Freudianism was still enormously influential, yet, as Catherine explained, Feldenkrais felt 'that one can talk and talk and talk and nothing changes, but if you pay attention to how you move it's something concrete that you can actually feel, and then you can examine your habits and patterns in your movements and develop choice and awareness, and that in itself will change how you act, feel and think in the world.'

She told me that often the starting point for clients is 'thinking, intellectualizing, analysing and talking because that's what they find supportive, then the sensing comes later. It can be exciting when they feel something for the first time. A typical response is "I never imagined that I could even feel that, let along have some choice as to how I do things." One of the most important things that Feldenkrais would say is, "If you don't know what you are doing, you can't do what you want." You need to notice in great detail. You need to be able to make these subtle distinctions in how you move, to know how you are moving in order to do something different. If you do it in an obvious way, which is how most body-work occurs, it doesn't have the same attention to the detail of what's going on. He left a legacy of something like two or three thousand movement lessons, which explore every aspect of human movement and functioning, from how you move your eyes to how the way that you move your tongue affects you.'

She described how restriction of neck movements, for example, results in restricted awareness, which may trigger ancient fears of attack, which will present as anxiety. She described her own experience. 'I remember when I first came across Feldenkrais the most valuable aspect for me at the time was that I realized that I always had been someone who was looking ahead to the next thing. When I was eating breakfast, I was thinking about lunch. It was very difficult. I always felt a sense of frustration, impatience and restlessness all the time. Feldenkrais was something that, for the first time, helped me to be in the present. I could be walking and feel my feet, and then I wasn't in my head. I had something to pay attention to in my body apart from the uncomfortable sensations of my emotions. When I tuned into my body before, all I felt was anxiety, but with Feldenkrais it took me into my joints and my bones, and that kind of attention was really helpful. It also helped me not to jump into an interpretation of my anxiety which would then escalate the anxiety, so I would be bouncing back and forth – "I feel anxious, it must be due to this – I must change that – Oh, I feel even more anxious then I must change that etc. etc." – constantly trying to change my external circumstances. The more you are in that kind of mind state the less accepting you are and life, of course, is much harder.'

She spoke of the felt rewards of the practice, a state that sounds rather like Flow, of full attention in the moment:[5] 'If my clients can get that experience, that reward – because learning to pay attention can sound rather dull – but the reward is huge and the pleasure that you can get from being more attentive and more present can be both horrific in that you notice things you may not want to notice, but also the best flow feeling ever. Feldenkrais was always emphasizing the importance of taking pleasure in how you are moving. He talks a lot about how when you are more attentive and you do know what you are doing, and you know how to do what you want, then you are able to be spontaneous rather than impulsive, and spontaneity has got a whole lot more pleasurable implications to it than being compulsive. To be fully in your senses and to be moving freely and enjoying is a real motivation just in itself. It is good to remember the positive, particularly in psychotherapy, which has quite a history of negativity bias, with a focus on problems rather than rewards.'

In a paradoxical kind of reversal of Feldenkrais's intuition that talk alone could never heal psychological problems, Catherine told me that it was the people who came to her with pain that had propelled her into training as a psychotherapist: 'I did feel that so much of what they were bringing had little to do with how they were using themselves, their structure or bodily organization, but a lot to do with their state; what was going on in their lives, what difficulties they were holding and what happened in their earlier history.'

Now bringing both the physical and the verbal together, she acknowledged that her earlier training in Feldenkrais method has been of enormous benefit in her subsequent training as a psychotherapist, particularly in the field of somatic attention. 'My new experience of working as a psychotherapist is that one of the strengths that Feldenkrais has given me is this ability to attend to the person I am with and at the same time pay attention to what is happening to myself, which I notice seems to be much harder for some people who don't have that background of somatic attention. They tend to dive into the person and their story and emotion in a close-focused determined way, and in the process lose themselves. I have a way of being able to access presence. When I am in a challenging situation I can notice when I have, as it were, left myself and I can come back into my body. I think that I noticed faster than my training peers when this happens. Feldenkrais doesn't market itself as a healing, more as a way of teaching – experiential learning. I think maybe the healing aspect is partly in attending to the person, and them attending to themselves, and at the same time you attending to yourself. They feel listened to and that kind of listening – that both of you are listening at the same time – is a rare thing. The two of you, and what is between you in what Martin Buber called the dialogic relationship, can provide the healing moment.'

To me this attention mirrors what was described in Chapter Four on emotional attention with regard both to early child development and to contemplative psychotherapy. Again reflecting earlier themes, Catherine spoke of working with young women with eating disorders: 'the attention that they need is not of the adult to adult kind; it's sometimes preverbal, so there is this sense that "Can I hold this person by being present to myself and present to them in whatever they bring?" Often that in itself seems to be more supportive than talking about stuff. I think probably their difficult feelings are related to a very early part of childhood, so to relate to them like an adult is not really appropriate at this stage. It was quite a revelation to me that actually by holding myself in my whole full presence, body presence, seems to be more supportive than talking and talking and delving into their process too much with words. My experience through Feldenkrais of learning to pay attention to myself, my body and how I'm moving means I'm able to bring myself deliberately into being present – a place where my thoughts calm down, my nervous system calms down and I'm able to attend to my clients in this way. I can be more present and using my sensory perception is a great way to be present.'

CHOREOGRAPHER Wayne McGregor has not only been creating works for the Royal Ballet at Covent Garden, ballet companies in San Francisco, Paris and Moscow and his own contemporary dance company Wayne McGregor Dance (formerly Random Dance), but has also deeply investigated this process of somatic creation. Working for years with a small interdisciplinary team of neuroscientists and academics in projects with titles such as 'Distributed Choreographic Cognition', 'Choreographic Thinking Tools' and 'Choreographic Language Agent', he has immersed himself thoroughly in research and exploration into the nature of creative learning, dance and movement. Some of the results of this work were presented in a recent exhibition at the Wellcome Collection in London. They have also given rise to a practical educational resource, a workbook, *Mind and Movement*, designed to be used in schools and learning environments to develop creative potential and enhance choreo-graphic skills. Much of this work has the aim of helping dancers evade habit in order to create new and original dance movement.

To catch Wayne in the midst of all these activities is not easy: our first meeting in London was cancelled due to an overrun of filming, and finally we spoke on Skype. When I asked him about the importance of attention in his work, he replied: 'I think it's essential. Working with these cognitive scientists, we have realized that it's even more central than at first we may have thought, because unpicking any habits cognitively is about being able to observe them in yourself if you can, and then to give you a whole range of techniques to shift attention to other aspects, so that you're not concentrating on any particular point. So that if somebody would have a dominancy in thinking visually, and would always build visual pictures, to get them to work in an acoustic loop, a cognitive loop that is primarily charged by an acoustic idea, is actually quite hard. But you need to know that your first preference is visual before you can actually shift the attention. Think about all those Buddhist practices of attention and mindfulness and sensing bodies and chakra shifts and Chi Kung – all of these practices are about that – are about centring your focus on a particular area that releases something in your body or makes new synaptic links or changes the way you think about the brain/body connection. That's why it has been so fascinating to work so closely with these scientists.'[6]

When I suggested that it sounded somewhat paradoxical to use attention to expose habit in order to evade habit, he agreed immediately: 'Yes, definitely. We always say that artmaking is instinctive, but actually so often artmaking becomes habitual and that's different from instinctive, and so what I guess we are trying to do is to prime the instincts in a different way with new sets of information that make you think outside the box and make you actually deliberately make different kinds of connections, and this fuels a kind of experience or emotional memory or feeling on the inside that makes you do something different, and that's what's exciting about it, that's what's interesting about it.'

I suggested that there must be difficulty in keeping this on the edge, as it were, so that you didn't just exchange one set of habits for another. Wayne answered that this was 'partly about having techniques – to be able to observe them yourself, others observing them, and also what you do once you feel you are in that place. It's almost like those 1960s oblique strategies where you get a card and it would tell you "Look at this afresh" or "Turn this upside down" or "Think sideways". It's almost like that cognitively that you recognize habits. Also it's not always about undoing habits because sometimes you want to use the habit. If sometimes I want somebody who works amazingly on the floor who has a real tacit understanding of gravity, I want to use that skill, I don't want them to be doing jumping, I want to use the thing that I know is really useful. So I think that part of the skill is also to allow the habits to be contained in the work as well as understanding where that happens and where you want it to be diverted, especially if you are stuck or blocked, or feel that the flow in the body or in the imagination isn't working. This is when shifts in the attention really help.'

He continued, as with so many of the other people I talked with, by bringing in the concept of time. 'What is interesting is the relationship with time. Because attention is also about giving yourself time. I have noticed a lot and been told a lot by the cognitive scientists that sometimes I would give a creative task and the dancers would respond to it very quickly but they hadn't actually processed it; they hadn't really given themselves the time to be able to do the thing that was required of them. So I think for me there is a massive relationship between how time is ordered within the body and within the context of the day, within your experience of time exchanging between other people, that is central to the practice of attention shifting, or even locking into attention. I think they are totally inseparable.

'I think the brain needs time to process, however it's connected. I think it has quick responses – the fight or flight responses, the primary responses that get you out of the situation quickly – and then it has whole layers of different kinds of responses which take a different kind of engagement, which actually demand you to be in your own head, recycling, challenging, testing, pushing, shifting ideas around so that you can see different corners of it, that then expose something else. I often find that if I am really stuck in the studio that actually if I shift my attention to something else and then sleep on it, the next day the problem has been resolved in some way, over that time of sleeping, that time of doing something else – it conspires in a way to give you another solution and I love that. I definitely think that there is a symbiotic relationship between time and attention, which is inseparable.'

When I asked Wayne if he thought that attention is something that can be trained, I inadvertently said 'intention' rather than 'attention'. He immediately responded both to the question and to my slip. 'I definitely think that you can, but it was interesting that you made that little Freudian slip of "intention" for "attention": there is something really interesting in the relationship between intention and attention. And actually I know that working with dancers sometimes you can get intentions through attention. As simple as something like, how is it that you shift attention through your body? How is it that I attend to my elbow? How I attend to my wrist or attend to the blood flowing inside my system as I am moving is a very different thing. Now I am building a different imaginative landscape that gives you a kind of intention. And that's different from, "Now feel something and let attention be expressed *through* the intention." There's lots of ways you can mediate between attention and intention, and you have to understand that they are different and then you can play them like an instrument. So, yes, they can be rehearsed, they can be refined, they can be practised: techniques can be put in place to be able to work more subtly with a range of options, so that you are not just moving between one and another in a really binary way, so there are sophistications in that learning and that really is analogous to playing a musical instrument. It's gaining techniques that allow you to be freer: it's not gaining techniques that bind you down. That's the point of technique. The point of technique is not to be rigid; the point of technique is to release you to express differently.'

In the exhibition at the Wellcome Collection, attention had been drawn to what was termed 'distributed cognition', the idea of a group understanding, which can enhance this kind of training. According to Wayne: 'It's always about transactions of energy between people, but it's understanding people's communicating, it's having an idea that propagates around a group, and elements change or are taken on in different ways. But first you have to prime where the attention is. Sometimes you don't have to do that – people can do it totally freely – but at other points, where everybody is zooming in, zoning in to a more particular attribute of something that becomes then very interesting. Phil Barnard [a neuroscientist who has done a lot of work with Wayne McGregor Dance] does this very interesting experiment with me where he says: "Well, you can either listen to the words when you are speaking, you can listen to the words that you are saying, or I can ask you to listen to the hiss in your voice and you will realize how much you hiss when you are speaking." So when I am talking to you now, you can hear that I have quite a lot of hiss in my voice [laughs]. That very small shift of attention allows you to experience something and understand something that you hadn't when you were just thinking about the words. I think if you can encourage dancers to be able to do that in a whole range of circumstances it just offers options creatively and that's super exciting. That's one of the beautiful things about those practices where attention actually reconnects you with your own body, where you actually do those exercises where you do a bodyscan from the feet to the top of the head and it makes you literally taller. It's not imagining that you are taller. It actually creates space in your vertebrae and your hips and your legs and your joints and at the end of it you feel that the volume of your body inhabits the space in a very different way. And that's the power of the body/mind connection to be able to breathe in a different way, whereas most of the day we are all so kind of compact, not thinking, not knowing. It's a kind of state of knowing, isn't it?'

Wayne has also been noted for his use of cutting-edge technology, both in training and in performance, though he says that he is not interested in technological application as decorative but as 'using technology as a way to prime a sense that you couldn't imagine on your own. Even in the very early days of digital cameras we used to do a lot of filming of dancers and then playing them in reverse, so shifting attention but being able to see it in the visual system and then learning it back. It's a level of detail and experience that you can't hold in your own head, but that once you've seen it you are able to scan backwards more easily, you are able to lock into that. We did early experiments with the SenseCam camera, a wearable camera, which IBM wanted to use at that time for club culture, taking pictures of your friends really randomly. Basically we tested it in the studios for a few weeks. It takes a picture randomly every few seconds and when you watch it back as a quick time film – your whole day in three minutes – it's amazing. When you watch it you notice all the things you didn't notice in your day. That's the first thing – how much you edited out of your day. You notice an incredible amount of information in the three minutes that has been lost, but is incredibly valuable to you. What they realized later was that this was amazing for Alzheimer's patients. So these kinds of techniques, snapped almost like key frames, watched back where your attention is different, improves memory in Alzheimer's patients by over 30 per cent. I find that really interesting; technology allowing you to experience an attentional thing in a different way that then you can capture valuable information from.'

It was nice to hear of technology being used in a benign fashion to offset the worries about 'outsourcing' our abilities that seem to be prevalent at the present time. Wayne also thought that some of the 3D computer games that kids are working with can help to build a spatial and praxic sense that far outstrips that which earlier generations experienced when playing football, for example, and that such experience, far from limiting, might actually encourage or develop some of our senses. 'But', he warned, 'they have to be done in parallel. They cannot be divorced from the body.'

Wayne told me that the next step in using such techniques to train dancers is to extend to the audience such enhanced ways of attending. 'Now we have gone on a journey around how we can work on our own attention and really develop techniques to help people to shift their attention imaginatively, to make different kinds of creative decisions and to be innovative in their thinking, not necessarily only in dance. How might we be able to prime audiences to be able to do that also? Because we all look through our own filters, we collect evidence to reinforce the view that we already know; we are not really attending to the performances when we watch. How might we be able to do that? I'm quite interested in this body-broadcasting thing at the moment. How you can take biometric data, which is not just steps, and not just how many calories you have burned, but how you can take the signals from the body from the inside and use that in some way to generate a relationship between audience and performance? Could you imagine a time when you would watch a performance in which the state of adrenaline would be measured in some way, and the collective state of adrenaline in the room would be measured, and that would drive a particular experience that you would see on stage and how would that be useful?'

Laughingly he suggested that 'It's almost like a lie detector for whether or not you like a performance in a really interesting way. I think there is something really interesting in that in relation to attention because you would attend differently. I do think you would attend differently.'

I had read in an earlier interview with Wayne that he wished for performances to have 'more poetry, more transcendence and to be more "other" than the programme notes could express'. So I asked if this attempt at some kind of transcendence, the presentation of something 'other' for both audience and performers, might spring from enhanced, open attention – if attention might have something to do with it.

'I think it does. I think we live in a very textural world and a world of a lot of concreteness and looking for solutions and we are very bound in our thinking and I think that one of the amazing things about performance, and particularly about dance and music, is if you can allow yourself to go there you get to a state of openness, you get to a state of a kind of ambiguity, you get to a state of kind of a lack of clarity – things are slightly out of focus – all of the things that the dance critics say shouldn't be in a performance which I think absolutely should be things in performance. And I think that the only way to do that is if you allow yourself, or train yourself or work out *how* to shift attention, how to shift your own attention, how to get into these new states of being or these new states of preparedness. How do you actually do that? And this I think is why we can learn so much from Buddhist practice, or why we can learn so much from Chi Kung – how energy and attention is shifted in the body in really extreme and interesting ways. That's why the arts are so interesting in this domain because you get an embodied sense that you really can't get in any other way with technology at the moment.'

His reference to the lack of clarity, the out-of-focus ambiguity of such states, reminded me of Iain McGilchrist's descriptions of 'soul' as a kind of attention, a state of experience that can't be met head on, that comes in from the periphery. But, fundamentally, Wayne said, 'I think it's all in the quest for meaning of some sort or meanings. I think we can't help but look for meaning in things and it's important. But we don't want always to look for the same meanings in things. I think that's the real power of art when artists and artworks can actually guide an audience to shift their attention and open up something emotionally for them that they have not been able to access before. Perhaps it's always been there but they have just not been able to access it. Dancing anyway has always been an attentional art form, if you think about all the practices in dance even when you are training – that whole idea of shifting attention and visualization, all that is really central to dancing – how you are able to even think about attention in relationship to someone else, which is a skill in itself, not just your own attention but almost a projected attention with someone else and how that transfer of energy works – between bodies rather than just of your own body – between rib cages rather than just your own rib cage and that's the sophistication, if you like, of dance practice, it has central to it this idea of shifting attention with other bodies.'

As with all the other media we have looked at, so attention reveals a fresh language – here of movement. When Wayne speaks of a dancer's 360-degree scope of awareness, her sensitivity to what her partner is doing behind her, out of her sight, and compares this three-dimensional awareness to our normal view in terms of the distinction between a face to face encounter and a screen conversation on Skype, I realize how two-dimensional is my own somatic awareness.

THIS IDEA OF embodied three-dimensional awareness came up later in a very different context when I was talking with Robert Silliman in California. He represents a different, perhaps earlier, though now rare, form of somatic attention – that of the hunter. I asked him how important he felt attention was in hunting, and he said that it was vital, 'first for being able to find game and then to know where other hunting partners or other people are. You have to have the observation to be sure for safety where everything is. You can also utilize people you don't even know – that you might see if you are paying attention – to have them assist you in hunting without them knowing it. It's kind of on a par with stealth and patience and being quiet. They go hand in hand. Some unsuccessful hunters can just wander and be patient, but if they don't have the observational skills and the ability to attend they will not succeed. Sometimes you just have the sense that something might be looking at you, and you turn and look and there it is. Sometimes if you are carrying binoculars, you just have that sense and you pull out the binoculars and just there you spot game or a person. I don't know if it is just like a sixth sense or if it's experience.'[7]

Maybe it is experience, as when I asked him if he thought that these attentional skills might be learned, he replied that he thought they must be: 'If someone doesn't have those attention skills, they might be successful sometimes but not consistently. They say in hunting that 90 per cent of the game is killed by 10 per cent of the hunters. That's why in early groups and tribes they had designated hunters who were the most accomplished and had that skill. Obviously it comes down to the ability to take the game, but you have also to be able to find it and get close enough to it, and to do that you have to understand what it is doing, so you can, for instance, utilize the wind. You can't walk downwind to most game, so you have to pay attention to the wind, to the terrain. Successful hunters learn this and learn it better and better, and that's why a lot of old people that hunt are better at it than young people. A lot of young people give up, and I think it is probably because it takes so much time. In the future there are not going to be so many hunters. I am sure my brothers and I learned so much from my father, and he learned from his father, my grandfather, who had learned from his father. We hear stories still when we go some place about what was successful before. You watch and you file that away in the memory bank and utilize it some time in the future. Today there are possibilities to hunt in different ways, which isn't quite in the nature of old-style hunting, where people go out with a guide and a guide may provide a great proportion of the skill, and someone may only have to make the final shot. You still have to know what you are doing and you still have to accomplish the kill but it's not the same as one-on-one man–animal hunting that was classically done.'

I asked Rob if his attentional skills honed in hunting had affected his attention in other ways. He believed that they had: 'You pay attention to things that are around you more. You kind of find your place on the planet – you carry it with you all the time. Obviously a hunter has to pay attention to the elements, wind, weather, where people are around you, and so you develop, maybe, an attention to people and things around you. When you see something moving fast it catches your attention more quickly. Animals find you by movement for the most part, so you learn to move more slowly and cautiously. If you watch you will see that some animals come and stand for, say, thirty minutes in one place before they take another step, whereas people are just bustling and moving all the time. They are not looking and planning ahead. On the ranch at home, before I step outside, I look around and see if I'm going to disturb quail or doves on the pond or rabbits. I don't just burst out, and then when I come out I'll take paths that are less disturbing. If the owls are out I don't want to go right under them, but I can stand in places where they will tolerate me. If I just open the door and rush into my barn, I will disturb the barn owls and they will then move out for the day, or for some hours, so I just find it easier for me to modify my behaviour and move as if it's a wild situation. I'm not in a hurry. And the animals do the same with me. There are a lot of species of animals that put a sentry out. The sentry is responsible for watching for danger and then they alert the rest of the group. Quail will always have a sentry sitting on the top of a fence post or a tree. If I vocalize to them they recognize me and I can walk by them slowly. Hunters say that predators look forwards, while prey look around. A pronghorn antelope's eyes are well at the sides of the head, and he can almost see right behind him, more than 300 degrees. So obviously they are prey, they have to be able to watch all the time. Hearing too is very important. I watch the ears of jackrabbits or deer, they will listen in two different ways and look in another and that's how they stay alive. They also have an unbelievable ability to smell. We, the predators, have a different sense. We are paying attention and are focused ahead of us, we don't have to be looking around all the time, but we have to use the wind and train all our senses. An accomplished hunter will smell or hear an animal before it can smell or hear him. Despite all the advantages of modern technology like scopes, a good hunter still has to move slowly and be patient.

'In hunting you have to move slowly and try to see game before it sees you move. When we hunt with a group like my family, brother, father and grandfather, we sometimes do what we call "still hunting". You move slowly, stop and look, move slowly, stop and look. It's like a deer would travel or a mountain lion would hunt. We would traverse miles and miles for hours. So we might start in the dark at five-thirty in the morning and not meet up until ten. We would know where we were going to go and as we travelled we would only intermittently see one person on one side of us, 50 to 100 yards or more away and as we travelled on at any given time we would know where a person should be and if you watched carefully enough they would appear there in moments. So hours later we would arrive somewhere and my father would point and say, "Where's your brother?" and you could point and say, "He's over there," and then he would show up. It was that attention to each other and feeling the balance. It's almost like having a dance partner and you have to do the same steps and end up at the same place at the same time and you develop that. From years and years of practice we can still go out and end up at the same place and do the same thing even in a place we haven't been before. We can say we are going to meet and describe the place and we will be there. People that aren't able to be there, they don't hunt with us any more. Some people you just have to tell to stand in one spot, and stay there and watch for game and we will come back and find them. They don't have that ability to pay attention to each other that much.'

Robert also felt that such skills had transferred into other areas of his life, such as his job as an engineer: 'I worked as a diver for five years, hard-hat. And we were in the dark for a lot of the time – a year and a half on a pipeline in San Francisco Bay, where you couldn't see your hand in front of your face. First of all you have to be at peace with yourself, and know you are going to be able to go down there without being claustrophobic or worrying about a shark or another event. So basically you have to have confidence. But once you are down there, you have to know which way you are facing at all times. It you start moving you have to be sure where you are moving. So you are aware of everything around you. As soon as you go in the water, you're aware of which way you will go, and you have an idea of how far you have gone and when you get to the pipeline or breakwater you have to know exactly where you are at any given time. You memorize what it's going to look like at the start, so that it's imprinted. I think a lot of it is just patience and attention, consciousness and planning. It's probably similar in other aspects of experience. In athletics you use attention and you have to know where everybody is going to be. I was a quarterback on a football team and even though you practise a play, at any given time you have to be aware of where everyone is, you have to pay attention to the defence over which you have no control. Athletics is probably the same kind of training situation as hunting, you have to develop skills that are important in the rest of your life too.'

LATER, I SPOKE with a practitioner of another form of sporting somatic attention. Sam Bleakley is a champion surfer whose involvement in the sport has taken him to many far-flung places to experience the ultimate wave. He is also a fine writer and the author of *Mindfulness and Surfing*, a book which – possibly unusually in the literature of the sport – references Heidegger and 'dwelling', Julien and 'vital nourishment', Derrida and Flow, as well as Buddhism and Taoism. From the very beginning Sam distinguishes the mindfulness of surfing from that of other forms. He is well aware of the debates about contemporary mindfulness discussed earlier, and states clearly that he believes current approaches to mindfulness can become mindless, ego-centred and personal, leading to withdrawal from the world rather than engagement. In contrast his book describes a journey from *ego* to *eco*. He feels this goes against the tide of Western development of techniques of focus on self and inward life, as a result of which we have what he describes as 'an egological surplus and an ecological crisis'.[8] His presentation of the mindfulness of surfing is based on an ecology-centred bodymindfulness that is paradoxically, a moving out of mind into the world, moving against the grain of inner-directed thought and reflection into an acute sense of what the environment demands of us – where winds, current, beach shapes, wave types and lunar tidal-movements meet. In this sense we move from 'egology' to ecology and we generate a 'bodymindfulness', locating ourselves in place and space.[9]

To achieve this he describes a journey of staying in the present, dissolving ego and immersion in nature in which

to 'perform' surfing is a first step, to 'think' with surfing is a second step, to let surfing think you and perform you, or to be shaped by the total environment, that we conveniently reduce to the act of 'surfing' is a more expansive step still.[10]

Taking this further, he believes that such mindfulness can connect out to 'the bigger bodymind of the community, to an extended and shared cognition moving 'from self to other, from contextualized persons to social relations, then to politics', suggesting finally that 'Mindfulness can be radicalized.'[11]

I caught up with Sam on Skype from his home in Cornwall, where he had just returned from two trips – to the Philippines and to Mauritius.[12] He explained that the perspective he had taken for this book was to bring to mindfulness ideas of ecoperception and the ways in which being engaged in an outdoor activity like surfing immediately facilitates close noticing of the environment. 'Obviously,' he told me, 'surfing naturally facilitates high levels of attention and close presence and very much acting and performing on instinct. And at the same time, like all activities, the mastery of something comes from practice – and the ability to respond quickly in the ocean as a surfer improves as you spend more and more time in it. The ten thousand hour rule is quite interesting because it becomes all these levels of engagement with the present, which kind of are a good representation of the spiritual practices that in mindfulness comes from Asian philosophies. What is interesting about surfing is that there are all these layers of attention and close-noting, from being very much immersed in the ocean and responding to the breaking wave and riding the wave, through to awareness of meteorology and weather patterns that can sometimes be forecast or looked at weeks in advance. So, for example, I can see the ocean from the window here and I am very aware of what the conditions are like right now, but most devoted surfers are also, both subconsciously and research-wise, aware of what the forecast conditions are going to be. So if you know that tomorrow or in one week there is going to be a particular wind direction and a particular size of swell, you are already starting to piece together whether you have got time, where you want to surf, what the tides are doing, what the sand movement has been doing at the surf break you want to ride, and so you have all these different layers which have all these completely different timescales and space scales, from the big macro ocean weather systems to the very micro in the present right now responding to the curve of the wave. Surfing is also quite involved in travel because most people have to travel to the coast to surf. Even if they live on the doorstep of a beach, they still have to walk or drive there, and because surfers are very interested in chasing better-quality waves they tend to travel at least once a year on a trip, perhaps to some-where where the water is warmer, like Indonesia, where the waves will probably be better than Europe. So it brings together a lot of activity-based engagement with the environment.

'It isn't so much going into the self, it is engaging with the needs of the wider community and environment and not just the physical environment but the need for cultural exchange in the social world. A lot of surfing is potentially invasive in that typically, the white traveller, who could be seen as the historical imperialist, goes to the poorer, for example African, coastline and does something that might seem frivolous to the fishing community who don't necessarily have the free time or money to buy a surfboard or engage in a leisure pursuit. Yet surfing transcends those boundaries through the shared experience. I wanted my mindfulness book to go beyond the typical inner meditative opportunities, that things like walking or beekeeping might facilitate, to consider how we need sometimes to come out of the mind into the mind of the planet, and how an activity like surfing can remind us that we can attune ourselves to the needs not just of the environment – sea level change or pollution, global warming – but also the demand for positive cultural exchange.'

Sam spoke of the beauty of surfing and the contrast between the exhilaration of the wave ride and the patience, timing and waiting: 'It's arriving at the coast, looking at the conditions, putting on your wet suit and going into the water and sitting and waiting for a wave and moving around, and the ride is sometimes only seconds, maybe ten seconds for a good ride. I suppose that martial arts might share those moments of space, patience and emptiness with very precise activities that are very accurate. And what happens is that even if a surfer doesn't become highly skilled in the performance of riding a wave, he or she definitely improves in their ability to read a surfing environment, to notice which wave to paddle for, to move at the right time and to pop up on their board and balance and ride the wave for as long as possible. It's appealing for people who want to escape from the Internet and the overabundance of media. It is an aesthetic adventure for people. It delivers adrenaline and it certainly would make the heart beat very fast, and it will also offer a low heartbeat post surf, so active surfers would be very healthy in the sense that they would be very calm and together and their bodies are also experiencing that incredible fluctuation from fear adrenaline, anticipation, expectation, to calm contentment, and I think that's really healthy physically and spiritually and creatively. The activity isn't known for a high level of creativity in terms of art and literature. That's possibly because it takes place in the countryside away from urban centres, whereas for example skateboarding, where the adrenaline is similar and movement based, is urban based and more celebrated by artists and musicians.'

While I think he is right with regard to twenty-first-century urban music, I think I might disagree with this on other levels. William Finnegan's *Barbarian Days* is currently enjoying much success, and I would also point to Daniel Duane's *Caught Inside* as surf literature, not to mention an earlier generation of West Coast American surfing music and film.[13] Indeed in Duane's book I found an articulation of several of the themes expressed in the literature of place and wildness discussed in a previous chapter. He notices 'how unsettling it is to discover how hard-won is real understanding of place, how much it demands stillness and time: *real* time, daily visitation'. Duane also writes of the 'appeal of the sea' as 'a commitment to the mundane unknowable, a daily dose of the wild'.[14]

Sam explained that in his book he talked about something called bodymindfulness, suggesting that 'perhaps another surfer-writer might have framed it in the classical form of inner mindfulness but because of my own interest in geography I wanted to frame it more outward-looking. I thought this would be a new perspective on the literature on mindfulness.' This reading of mindfulness as a bodymindfulness, a within-body reflection as well as a within-mind reflection, is intended, he suggests, to recover the instinct of what he terms 'reflexivity-within-action', which sounds very much like the merging of action and awareness in Flow. In the book he describes how he thinks in music there is nothing closer to surfing than jazz and, particularly, drumming, where the rule is within a general frame of preparedness, to stick with the 'now'. When the drums take over the drummer, new improvised patterns appear that can surprise player as much as audience, just as Sam Richards and Lona Kozik described: 'You don't play what you want to play, you play what the music wants.'[15] The here and now of the sea perhaps merges with the hear and now of the music.

10

Attentive and Experiential
In-conclusions

Attention is the natural prayer of the soul.

MALEBRANCHE[1]

It has been extraordinary how themes and ideas have recurred and echoed again and again through the pages above. In conversations and readings, an observation encountered earlier would reappear in another context, slightly reshaped by its new surroundings, but surely pointing back or forwards to other discussions in the endless exchange about attention. Pairs and paradoxes abound: focused attention or free-ranging awareness, creative freedom or constraint, humility or conceit, stilling of self or expressing self, listening or speaking, speaking or saying, representation or presentation. Always the deliberation has been 'under way', in process, a movement akin to or resembling the movement of attention itself working itself out in action; in the solitary action of the creative or attentive mind and in the joint attention with others that is compassion – feeling with. It is a movement that is never finished, never finalized, a movement full of paradox – of the necessity of submission to find freedom, of the need of others to find the self, an endless dance between focus and extension, self and other, freedom and limitation. Such paradoxical pairings are echoed again in the descriptive duality between what is happening at the neuronal level and the living felt experience. To consider attention is like the eye attempting to see itself. Yet in such consideration, an attention to the dancing itself rather than a search for a fixed form, a sense of wholeness and richness, if not of finality, may be experienced.

The more I have written, the more people with whom I have spoken, the more I have realized that the subject of attention exceeds any attempt to encompass it. So much has been unattended. I have said little about the dangers – and indeed the potential – of social media. Others have and will continue to address this important subject. My hope is only that I may have inspired some awareness and curiosity about the process and promise of our attention. Thus, bearing in mind an understanding of its ultimate in-conclusion as without closure, I would like to attempt to draw together some tentative threads from all that has been written above, with particular reference to the three themes I outlined at the start. Even as I write this I notice a shift in intention – from open attention to a search for certainty, clarity, meaning and the imposition of pattern. Surely a human impulse, but one that I suggest should be resisted, or at the very least noted. The need for knowledge may impose premature closure. To appreciate the conflict – closure and certainty versus openness, curiosity, being at peace with what is – and to stay just a little longer with open attention, resisting the lure of the known pattern, of the habitual response, may change the pattern and open up new paths and new horizons. Thus inconclusion, but also some threads – threads that may be woven into a fresh tapestry, new paths that may open up into a new clearing, always changeable, always temporary, expressions of time and impermanence and mutability. I hope this has made you consider, appreciate and take the time to attend to the wider horizons of attention, beyond the task or object at hand. I hope I may have succeeded in showing attention as a skill, a practice, which can be cultivated. All the various participants in the conversation would seem to agree that it is a trainable competence, one that can be practised in many and various ways, and which brings rich reward. They have also pointed towards some practices that may be particularly beneficial to good life, and how these might be encouraged and shared.

The third theme is the idea of enhanced experience as a contemporary secular 'religious' experience – the 'everyday sublime'. Adherence to organized religions is declining in the West, yet every poll concerned with 'soul' or 'spirituality' reveals a desire for some kind of meaning, transcendence or peak experience. I want to suggest that it may be through practices of attention that we may achieve these. Practices of attention provide the pathways to deepened, meaningful experiences, as evidenced by the long history of art, meditation and prayer. There are difficulties of terminology here: 'religious', 'spiritual' and 'secular' are terms that appear and reappear in any discussion of these topics. Dale S. Wright, professor of religious studies at Occidental University, Los Angeles, has recently written an excellent article opposing the use of 'secular' in a description of contemporary Western Buddhism. He bases his argument on the understanding that 'secular' fails to include sufficient recognition of 'the great matters' of birth and death, which are

> our historical assignment, our calling: to affirm the religious dimension of human life by re-envisioning and reformulating spiritual sensibilities at the cutting edge of contemporary thought, practice, and experience. In contrast to the triumphalism and dogmatism that characterize both theism and atheism, a thoroughly post-theistic religious practice would be experimental, moving forward in humility and openness toward a range of new possibilities for enlightened human life.
>
> This post-secular orientation would begin in contemplative practice in order to build habits of mind and body inclined to question the instrumental character of current common sense, which imagines ever new means but never ends, leaving us without ideals and goals suitable to our time.[2]

Korean artist and writer Jungu Yoon, in his exploration of spirituality in contemporary art, uses the term 'numinous' to describe a concern with similar matters, evading the traps of 'spiritual' and 'religious'.[3] Yet both Wright and Yoon and those, like Stephen Batchelor, who stick to 'secular', are equally concerned

a1

with the great matters. I believe that if we focus on practice, in Wright's words, 'the quest for authenticity, for wholeness, for self-actualization and self-realization' and engage goodwill and the recognition that human beings and human cultures are always underway and never completed, a helpful and healthy engagement is possible despite differences in terminology. Radical attention reveals interdependence, contingency and emptiness and should encourage us to be tolerant of unknowing, dodging the desire for certainty, essence and identity. It engages that negative capability of John Keats, 'being in uncertainties, mysteries, doubts, without any irritable reaching after fact and reason',[4] which is echoed today by MacFarlane's 'awareness of ignorance', and Matthew Crawford's conversational approach to knowledge in the stochastic arts.

We have explored how some artists pay attention and how their journeys into attention may take us by ear, eye and hand, to follow in their steps through word, sound, image or movement. We have seen how meditation and mindfulness can take us on the same journey, and how sport and craft provide other routes to hone our skills of attention. What threads run through these different approaches? What are the ingredients for this kind of enhanced, meaningful and enlightening experience? There are, I think, steps on the path. First a letting go, or at least a deep questioning, of our usual understanding of words, then of thoughts, and then of certainty and self; after that, an opening up, an embrace of what remains, an open, equanimous, indifferent acceptance. These are the steps of *practice*, a practice that may alter our habits; that could change our everyday default mode.

Letting Go of the Words

We can lose life in the narrative. T. S. Eliot suggested once that 'We had the experience but missed the meaning.'[5] But we can also lose the experience within elaborations of meaning, let theory envelop the lived experience and, as the great mythologer Joseph Campbell put it, 'let the concept swallow up the percept . . . thus defending ourselves from experience'.[6] Slowly, subtly, the abstraction, the story, becomes more central, more real than the felt experience. The name and its habitual expectation constrict the experience. As Samuel Beckett noted, 'No need of a story, a story is not compulsory, just a life, that's the mistake I made, one of the mistakes, to have wanted a story for myself whereas life alone is enough'.[7] Maybe we do need both, but how easy and unnoticeable is it to lose the latter in the needs of the former? Image takes over from reality, representation from presence, and what we wish something to be from what perhaps it is.

Here we meet again the distinction Heidegger made between *saying* and speech, and Henry Bortoft between attention to what is experienced and the *experiencing* of what is experienced. Paying radical attention to language, we may not entirely lose the words, but we may use them differently. Not least, we may attentively notice the way the grammatical inevitability of our language tends to turn action and verbs into unmoving nouns, and consider the habitual effect this has on the way we experience.

Perhaps the most radical verbal attempts to take us beyond language are to be found in the koans of Chan and Zen Buddhism that David Hinton has beautifully described as 'paradoxical scraps of story and poem that tease the mind outside normal linguistic structures into profound nonverbal depths'. The purpose of this is to enable us to pass through the gateless gate, which stands as the title of one of the koan collections. It is the 'gateway between subjective and objective, consciousness and landscape, human and stellar'. The gateway is gateless because in our original nature, beyond the gates of identity and language, 'consciousness and landscape and stars are all of the same tissue'.[8]

Letting Go of Thoughts

This we have seen particularly in the formal practices of meditation and mindfulness. Their central message is that of the laying down of thought, name and story, the reining in of the wandering mind for a limited period, in the hope that such formal practice may instil permanent benefits in training the unruly thoughts and revealing the awareness beneath. Bringing ourselves back more fully to our embodied experience in the present moment first reveals and then may tame our identification with, and subservience to, our wandering thoughts. It may lead us back to an appreciation both of our senses and their interface with the world.

Attention is linked to self-control, to emotional self-regulation, as demonstrated in developmental studies of children. Understanding this and in fear of its loss, in 1772 Dr Johnson wrote

> I had formerly great command of my attention, and what I did not like I could forebear to think. But of this power which is of the highest importance to the tranquility of life, I have for some time been so much exhausted.[9]

Losing our power of attention, we fall into the power of our thoughts and emotional self-narrative. Attending to them allows us to see that they are just part of a greater picture of sensory embodiment and enworldedness, which, paradoxically, may lead us to disidentification and to a widened awareness that deprives thoughts of their habitual hegemony.

Our thoughts are often used in the service of certainty that is actually ungrounded, without foundation. Wider awareness beyond self gives us a true ground. All the people with whom I have spoken in the process of writing this book have spoken, in some form or another, of uncertainty and openness to whatever arises without expectation or fear; of what I might term a 'creative humility' in the face of experience.

Letting Go of the 'Self'

Here, first, a story:

One day I was riding down the lane beyond my house. It was a cold winter's day with bright sunshine after a freezing night. Coming round a bend in the track suddenly the winter sunlight was shining through the frosted hedgerow. Each frost crystal became a diamond and the track a jewelled tunnel shining in the sun. It was breath-taking. As my horse walked through this glory, I turned to try and catch a last glimpse, but then I was facing in the other direction, the sun was behind me and I between sun and crystal. All that remained was a damp green brown hedgerow, winter bare, no sparkle, all glory gone. I have never forgotten that moment. It showed how dependent we are on where we are looking from. It was the same hedge in the same landscape but it was a different experience. Only much later did I realize that the wonder faded as I changed my position in such a way that I was in the way, between light and object. That seems symbolic. It is so often the I that gets in the way.

When the I blocks the view it takes over the centre of everything and everything is then seen from the perspective of this subject, thus becoming only object. A loosening of this stance may allow for a different reception of the world, the emphasis being on recep-tion rather than imposition of self and its self-centred view. The poet Shelley aspired to the dissolution of the self, wanting to be as an Aeolian harp so that the wind could play on the receptive strings of his body and mind, writing 'Make me thy lyre'. When writing to his friend Leigh Hunt, requesting that he publish one of his poems anonymously, he said, 'So much for self, self, that burr which will stick to one. I can't get it off yet.'[10] We have seen how both Buddhists and neuroscientists describe the self very differ-ently from the way we usually conceive of it as a singular identity: 'If there were a solid, really existing self hidden in, or behind the aggregates, its unchangeableness would prevent any experience from occurring; its static nature would make the constantly arising and subsiding of experience come to a screeching halt.'[11] They see the self as described by writer and neuropsychologist Paul Broks, as 'divided and discontinuous'. In summer 2016 Broks broadcast a series of radio programmes based on case histories that deeply questioned our usual understanding of the self as a permanent, finished identity. Such a conceptual shift also encourages an emotional shift – to a more humble, more participatory engagement with world and with others. If our I is seen as more compounded, more changeable, it is also more implicated in its interaction with what is not-I. If this understanding can be achieved, without the endless desirous grasping of what will strengthen what we see as our identity, and fearful pushing away of whatever might endanger its (fictional) solidity, our experience will be radically enlarged – even perhaps en-lightened. For, as Annie Dillard so beautifully described, joining together the loss of words and self,

> The death of the self of which the great writers speak is no violent act. It is merely the joining of the great rock heart of the earth in its roll. It is merely the slow cessation of the will's sprints and the intellect's chatter: It is waiting like a hollow bell with stilled tongue.[12]

Embodiment

Letting go of thought, words and self-image does not leave us afloat and disorientated, because it is accompanied by embodied attention. As Dillard encourages, 'Lick a finger: feel the now.'[13] So many of the participants in this book refer to body practices and body awareness. Attention is closely linked to breath. We tend to notice both attention and breathing in default of normal function; when our attention is tightly gripped or lost in reverie and when breath is overexerted or choking. Correspondingly almost all practices of attention begin by placing attention onto breath, bringing body and mind together.

This embodied attention, together with the loosening of word, thought and self, may enable us to move from place to ground, as in Stephen Batchelor's differentiation of ground and place. Place stands for all that we commonly identify with, our location, social status, religious and political beliefs and our belief in being an independent ego. It is where we commonly 'take a stand' to defend our self. Ground, in contrast, 'is the contingent, transient, ambiguous, unpredictable, fascinating and terrifying ground called "life".'[14] Thus it is a groundless ground.

Antony Gormley, who has based his entire working practice around the experience of embodiment, making sculpture that he feels can become a reflexive instrument rather than a representational narrative, credits Buddhist sculpture with teaching him about making sculpture that is about being rather than doing: 'There is an important distinction to be made between a sculpted body used for the illustration of a narrative and the making of a body as a still object that invites projection.'[15] He speaks of such works as both coming out of concentration and demanding a form of concentration, describing the resulting sculptures as 'a catalyst for reflection'. He writes of two paths in his work: one to make objective, human-scale models of the space of a lived body, the other to make 'enterable structures that allow the individual to experience consciousness in a new way'.[16]

Edmund de Waal expressed anxiety about our increasing ignoring of tactility; others have spoken about the 'outsourcing' of embodied human responses. What we may be in danger of losing is the enrichment of experience.

This grounding in the body ripples out into the next heading that is about participation in environment and context. Full understanding of embodiment requires a somewhat radical departure from our usual ways of thinking. In this new understanding of embodiment, we no longer say we *have* a body, but we *are* a body. Rippling out from this are challenges to many of the still commonly uncontested dualities such as the Cartesian mind/body division, the split between rational cognition and emotion, self and other and even the division between self and world. In turn, this throws into question all the value judgements that have arisen from these long-held dualistic views: that cognitive intelligence is of more value than emotion; that brain-based work is superior to hands-on practices. In the words of one of the clearest adherents of this body-centred view, Guy Claxton, whom I spoke with on the subject of education:

> The entire human system is self-organizing. There is no 'little person in the head' who does the cognitive heavy lifting; who 'pays attention', 'makes decisions' and 'plans actions'.
>
> We are a series of nested or unfurling systems, and to speak of having a body is a misrepresentation of the system, driving a wedge between body and so-called 'owner', when really 'the 'I' is as much an upwelling from the interior as the thought or the action itself. I am a body–mind context constellation, ever changing and ever willing up.[17]

And this body–mind constellation is also enworlded.

Embracing What Is and What Is Not: Participation

The first three headings involve a letting go. For what purpose? Is it that we may better, more openly come to what is – embodiment and world? Freed from narrative, concept and self, the lynchpin of both story and thought, reunited with our flesh, we may perhaps attend to world in a different manner – a manner that is more open, more innocent and more participatory. Paradoxically, letting go of words and of self has the effect of uniting us, mind with body, our being with world. Iain McGilchrist suggests 'that we neither discover an objective reality nor invent a subjective reality, but that there is a process of responsive evocation, the world "calling forth" something in me that in turn "calls forth" something in the world'.[18]

It was the view in ancient China, is that of the Enactive school of cognitive or neuroscience and philosophy of mind and of phenomenology, and also is to be found in the experience of indigenous people. Barry Lopez, writing of events and encounters when travelling with indigenous peoples, clearly articulated this perspective. He described how, expanded beyond his own parsing of an event 'into our customary divisions of time and subject and object, my friends had situated themselves within a dynamic event'. Their approach included 'an expansion of temporal boundaries' and an attention to the patterns in what they encountered rather than to isolated objects. Thus they allowed and fully experienced the unfolding of an event. 'Unlike me,' he says, 'they felt no immediate need to resolve it into meaning. Their approach was to let it continue to unfold. To notice everything and to let whatever significance was there emerge in its own time.' So they allowed the unfolding of the event and experienced it as full presence, in contrast to his own more superficial experience, thinking about it, and collapsing 'mystery into language'. As he describes it, 'For me, the bear was a noun, the subject of a sentence; for them, it was a verb, the gerund "bearing".'[19] As McGilchrist stated, their world view arose from their attention.[20]

This letting go, this altered relationship to language, self and experience has been carefully described (in more Western philosophic and less embodied terms) by Henry Bortoft as the way in which 'something in the world [that has not appeared] evokes a response [in the perceiver] which calls forth that in the world which evokes the response [appears]'.[21] The process is a dynamic whole that knits together being and perceiving, in the same way that Heidegger sought: 'Being means appearing. Appearing is not something subsequent that sometimes happens to being. Being presences *as* appearing.'[22] If we allow and understand that something in the world evokes a response in this way we avoid the dichotomy of being and appearance; they are somehow neither the same nor different; there is distinction but not separation; and that paradoxical participatory space, allowing for toggling between being and appearance, makes possible intensity and grandeur. This requires a *shift of attention* within experience; a shift that takes us upstream into process rather than our habitual focus on the downstream product. That is to say, upstream from what is said into the saying; upstream from what is understood into the event of understanding. So doing, the event of understanding becomes the appearing of meaning, and language becomes the event of disclosure. This is what Heidegger pointed to when he said that 'the essential being of language is *Saying as Showing*'.[23] The barrier to proper understanding of language and world, as Bortoft points out, 'is what can be called the "myth of subjectivity", which emphasizes self-consciousness and consequently conceives the individual self as being the centre and origin to which everything must be referred.'[24]

Maurice Merleau-Ponty had earlier pointed to Cézanne as illustrating the phenomenological stance, quoting the painter: 'A minute in the world's life passes! To paint it in its reality! and forget everything for that. To become that minute, be the sensitive plate . . .'[25]

This requires that we hold lightly and question our habitual views and performance of self and language. Our routine ways of being in the world are that of forgetfulness of the creativity involved in the relationship of self and world, a falling-off into a habitual, unreflective reaction, as our attention is focused on the appearance of what has already appeared, structured by the subject/object divide, rather than being alive to and participating in the appearing. If we can shift our attention upstream to the process of appearing, the process of *Saying* that is disclosure, our relationship with the world changes, is re-visioned.

A recent article from the *New York Times* by Lawrence Berger speaks of this distinction. It was called 'Being There: Heidegger on Why Our Presence Matters'. The author attempts to take the subject of attention beyond the realm of cognitive science into what we could call the felt sense:

attention has to do with all possible modes of human existence – all senses (visual, auditory, etc.) and other modalities such
as thought, emotion and the imagination. Any information
processing that provides access to things so they can be
represented must first go through the filter of attention.[26]

For Heidegger, attention is the way that things come into presence
for us, and therefore staying with, and paying better attention to,
an entity thus enables a deeper revelation of its nature. Such attention, which we have seen illustrated in the poetry of Rilke, the
painting of Cézanne and the philosophical writing of Heidegger
and Bortoft, allows for a different relationship with the world – a
participatory relationship, in which the being of the object of our
attention, and our relationship with it, cannot be conceived inde-
pendently of the entire context in which we are engaged. In the
language of Heidegger, we 'belong together' with the objects of
our attention. 'We are made manifest together. Rather than being
discrete entities, the relation comes first, and to the extent to which
we are related matter for what we and the stone (the object) ultimately *are*.'[27] And such attention matters. 'On one view we are
fundamentally cut off from the world, while on the other we are
in direct and potentially profound relation with the people and
things that we encounter.'
This reflects the more embodied illustration given by Barry Lopez
in respect of relationship with place.

The determination to know a particular place, in my experience, is consistently rewarded. And every natural place, to
my mind, is open to being known. And somewhere in this
process a person begins to sense that they *themselves* are
becoming known, so that when they are absent from that
place they know that place misses them. And this reciprocity, to know and be known, reinforces a sense that one is
necessary in the world.[28]

Time

In almost every conversation I have had during the writing of this book, attention has, in some way or another, been linked to questions of time: taking time, slowing down, setting in a greater timescale. Wayne McGregor spoke of the 'symbiotic relationship' between attention and time. Hunter Robert Silliman and potter Edmund de Waal spoke of the hours necessary to instil a skill. Writers of place speak of the time necessary for attention and attendance to their places, Wendell Berry to his farm, Barry Lopez to a wider environment, Daniel Duane to his surfing spot. Feldenkrais method is founded on the experience of slowness. Almost everyone I spoke with mentioned slowing down, taking time. We live in an age of speed, are encouraged at every turn to 'save time', 'take a short cut', 'pack more into our day'. Worryingly, this also applies to our children. The empty time, the space for reflection, the invitation to imaginative play is encroached upon by provided and professionally produced entertainment. I surprise myself by thinking today of the boredom of which I complained in my childhood as something to be desired for today's children.

Time is such a vast subject, is so inextricably and somewhat mysteriously entangled with our experience that to do justice to it would call for a book in itself, calling for far more space and care than I can spare here. I will limit myself to noting its importance, and that attention requires time, perhaps is even inseparable from time. Certainly attention can change our experience of time; taking time may encourage and enable the following heading. For, as Emily Dickinson told us, 'Forever – is composed of Nows.'[29]

Epiphanies of the Everyday

Stephen Batchelor often uses the term 'everyday sublime' to describe a different, enhanced way of seeing and engaging in everyday reality that he speaks and writes about in relation both to meditation and, as explored in an earlier chapter, to art practice. He suggests that there is an aesthetic dimension to insight as well as affective and cognitive dimensions, stating, 'Meditation originates and culminates in the everyday sublime.' He has no interest in the recitation of mantras, or the visualization of Buddhas and mandalas, in gaining out-of-body experiences, lucid dreaming or parapsychological experiences. To him, meditation is about the ability to embrace just what is happening to this organism in this moment:

> I do not reject the experience of the mystical, I only reject the view that the mystical is concealed behind what is merely apparent, that it is anything other than what is occurring in time and space right now. The mystical does not transcend the world but saturates it.[30]

To get to this understanding, in the sense that the sublime stands for that which exceeds our capacity to represent, he points to the fact that, properly, contemplatively, seen, 'The world is excessive, every blade of grass, every ray of sun, every falling leaf is excessive.'[31] All are beyond adequate capture in word, image or concept. (Here I think we can perhaps add that Heidegger would point to *Saying* and Bortoft to an 'upstream' understanding as evading such inadequacy.) Such sublimity, he states, 'brings the thinking, calculating mind to a stop, leaving one speechless, overwhelmed with either wonder or terror'.[32] Such wonder or terror is to be found only on the other side of speech, of calculation, of control from our self-centre. Lying on one's back in the dark, staring up at the immensity of the night sky and the deep dark depths of the stars, may bring on a kind of wondrous, terrifying overwhelm. It is too much for our everyday selves to comprehend, which is why we habitually retreat. We draw back from the vastness and view it from the restricted, safe perspective of our habitual selves and encompass it with our petty desires and fears and categories. Yet the everyday sublime is there; it is our ordinary life as viewed from a perspective that is able to embrace the totality of our condition. It is not confined to the extraordinary. As James Turrell said to me, 'We've forgotten about small things and how dear they can be.' Wittgenstein wrote, 'It is not *how* things are in the world that is mystical, but *that* it exists.' We must let go of our habitual thoughts and defences, acknowledge the possibility of freedom from them, and engage in practices that both reveal this everyday sublimity and emerge from its immensity. McGilchrist has suggested that 'soul' is perhaps a quality of attention.[33]

Bringing together art, ethical and spiritual practice, Batchelor suggests that greed, hatred and ignorance are an-aesthetic: they dull our perceptions, presenting a life rendered flat.[34] The arts may thus be a practice of embracing the totality of life, in all its tragic and joyful aspects, revealing its everyday sublimity if one is particularly alive to its aesthetic dimension. Matthew Crawford drew attention to this in his 'erotics of attention'. A practice of radical attention can lead to a way of life that is responsive, 'empathetically, ethically and creatively' to our situation, which is the ever-changing, tragic and joyful, empty experience of being here.[35] It is, as Batchelor says a lifelong practice. In this way attention comes to be, as Malebranche suggested, a natural kind of prayer.[36] As W. H. Auden wrote,

> To pray is to pay attention to something or someone other than oneself. Whenever a man so concentrates his attention – on a landscape, a poem, a geometrical problem, an idol, or the True God – that he completely forgets his own ego and desires, he is praying.[37]

He went on to say that 'the primary task of the schoolteacher is to teach children in a secular context, the nature of prayer.' Kathleen Jamie, poet and writer, describes a time when her husband was critically ill:

I had not prayed. But I had noticed, the cobwebs, and the shoaling light, and the way the doctor listened, and the flecked tweed of her skirt, and the speckled bird and the sickle-cell man's slim feet. Isn't that a kind of prayer? The care and maintenance of the web of our noticing, the paying heed?[38]

Emptiness and Openness, Uncertainty and Paradox

I wrote earlier about philosophies of emptiness, where I found understanding them to necessitate a shift of attention, initially similar to a figure/ground reversal from our Western default setting on substance and presence. In aiding that readjustment, I am deeply aware of several recent writings in the West coming from an appreciation of emptiness, ambiguity, uncertainty, not knowing and even what appears to be 'negative principle'. Many with whom I have spoken have drawn attention to such themes: Iain McGilchrist speaking of periphery and metaphor; Robert MacFarlane writing of the 'decentred eye', 'centreless nature' and 'awareness of ignorance'; Jane Hirshfield's reference to the unsayable and impossible; and, earlier in time, Keats's expression of negative capability. Matthew Crawford emphasized the importance of constraint, and described 'attentional commons' as a 'negative principle'. These matters come up in interesting recent work by scientist Terrence Deacon, who is locating the arising of both life and mind in an emergent dynamic approach that takes into account both what is present and what is absent, finding a capacity for work in 'specific absent tendencies, dynamical constraints that are critically relevant to the causal fabric of the work and are the crucial mediators of non-spontaneous change'.[39] Deacon makes an analogy between the challenges posed by the mathematics of zero and those posed by what he calls the 'ententional' properties of living and mental processes. *Ententional* is an adjective to describe 'all phenomena that are intrinsically incomplete in the sense of being in relationship to, constituted by, or organized to achieve something non-intrinsic.

This includes function, information, meaning, reference, representation, agency, purpose, sentience and value.'[40] He says that modern science, in the way the Middle Ages attempted to exclude a role for zero, has attempted to exclude the mark of absence as a factor in justifiable explanation. In contrast, he has attempted to undertake what he too calls a figure/ground reversal to reconceptualize 'some of the most basic physical processes in terms of the concept of constraint: properties and degrees of freedom not actualized'.[41] He has coined the term '*absential*' for these constraints. Absential stands for 'the paradoxical intrinsic property of existing with respect to something missing, separate and possibly non-existent. Although this property is irrelevant when it comes to inanimate things, it is a defining quality of mind.' He describes it as a 'constitutive absence', that is a critical defining attribute of ententional phenomena.[42] His book ends,

> By rethinking the frame of the natural sciences in a way that has the metaphysical sophistication to integrate the realm of absential phenomena as we experience them, I believe that we can chart an alternative route out of the current existential crisis of the age – a route that neither requires believing in magic nor engaging in the subterfuge of ultimate self-doubt. The universe *is* larger than just that which we can see, and touch, or manipulate with out hands of our cyclotrons. There is more here than stuff. There is how this stuff is organized and related to other stuff . . . It's time to recognize that there is room for meaning, purpose, and value in the fabric of physical explanations, because these phenomena effectively occupy the absences that differentiate and interrelate the world that is physically present.[43]

In the face of all of this, to ignore constraint, contingency and emptiness, to emphasize our freedom would seem to be self-defeating. For Crawford certainly, the answer lies in skilled practice and careful attention aware of its tacit dimensions of embodiment, engagement and dependence on other, on tradition, culture and context. Interestingly, though in more abstract terms perhaps, Deacon considers the most important part of his theory to be the idea of the capacity to do work.

Iain McGilchrist suggests that 'in creation "no: comes first'. He cites both Lurianic Kabbalah and neuroscience, to which I would add Taoist thought. In a description that also illuminates ideas of presencing and representation, he uses science to underpin phenomenology, stating that the primacy of no

> appears to intuit the way in which I believe everything comes into being for us phenomenologically, through the interaction of the hemispheres. The first 'act' is not the making of something happen, but an open receptive attentiveness by the right hemisphere in which all new experience begins. It creates a space for something to be. What is received is then 'poured' into the various categories and systems that the left hemisphere brings to bear on it, but these are inadequate to contain the meaning that was there in the first manifestation or 'presencing', and they break down. The meaning has to be returned to the right hemisphere to be 'restored' by understanding it as a whole again.[44]

So it is not only Buddhists who seek an acceptance of contingency and a middle path between the peaks of paradox. While writing this I came across some wonderful statements about paradox. In a series of lectures that attempts to address both what is evident to science and that remainder that escapes it, Iain McGilchrist noted that 'at the heart of every wisdom tradition there is paradox'. This was accompanied by a wonderful positive description of paradox as 'a symbol of resistance, a warning that thinking does not match reality, and that therefore reality is not just what we think'. It was accompanied by a quotation from physicist Niels

Bohr: 'How wonderful that we have met with a paradox. Now we have some hope of making progress.'[45] Apparently Bohr chose the yang/yin symbol and the words CONTRARIA SUNT COMPLEMENTA (CONTRARIES ARE COMPLEMENTARY) for his coat of arms when honoured by the Danish government. Kierkegaard said:

> One must not think ill of the paradox, the thinker without a paradox is like the lover without passion: a mediocre fellow . . . the ultimate paradox of thought: to want to discover something that thought itself can't think.[46]

Writer Annie Dillard, one of those who have best described the transfigured experience of the everyday sublime, expresses the paradox of the enhanced attention that is necessary. She writes of

> another kind of seeing that involves a letting go. When I see this way I sway transfixed and emptied . . . But I can't go out and try to see this way. I'll fail, I'll go mad. All I can do is try to gag the commentator, to hush the noise of useless interior babble that keeps me from seeing . . . The effort is really a discipline requiring a lifetime of dedicated struggle; it marks the literature of saint and monks of every order East and West, under every rule and no rule, discalced and shod. The world's spiritual geniuses seem to discover universally that the mind's muddy river, this ceaseless flow of trivia and trash, cannot be dammed, and that trying to dam it is a waste of effort that might lead to madness. Instead you must allow the muddy river to flow unheeded in the dim channels of consciousness; you raise your sights; you look along it, mildly, acknowledging its presence without interest and gazing beyond it into the realm of the real where subjects and objects act and rest purely, without utterance.[47]

Practice

All the above thoughts have hopefully demonstrated the importance of attention, and the rewards of enhanced awareness.

Unfortunately, this tolerance of paradox and these ideas of emptiness and openness, of presence and absence, self and world have not been, are not, the common ways we consider and experience selves, world and time in the contemporary West. That more ordinary manner is still marked by Cartesian dualities and a belief in a solid identity, an exterior world and a regulated time. On the other hand, phenomenology, Taoism, Buddhism, the Enactive wing of neuroscience with its emphasis on the 4E cognition – enacted, embodied, embedded and extended – and also many thinkers and artists all point to a more participatory view of the world in the manners described above. They emphasize the power of attention, through openness and receptivity to the occurrences of the moment, to reveal the complex, often paradoxical dance between experience and world, self and other. For Buddhists, carefully honed attention reveals impermanence and insubstantiality, the compounded process of selfing and being and its reliance upon expectation and disposition. For scientists, interoceptive and exteroceptive attentional skills ensure that more of our somatically embodied processes are available for the best possible response to our environment. For artists, honed attention reveals the everyday sublime and aspects of the world afresh.

Careful attention to the present moment takes us into a different understanding, an embodying of our selves, which reveals our participation in the world, a world enriched, seen not merely as resource but as source. A receptive reflective awareness is an intentional state of openness to whatever arises. Another of the commonalities revealed by the conversations of this book is the idea that the achievement of such a state is a skill that can be practised. As Guy Claxton says, 'Attention is like a muscle that becomes stronger and more controllable over time; apparently it can be developed just like any other habit.'[48] In the words of Daniel Siegel, this 'creates a flexibility in self-regulation that may enable an individual to profoundly shift out of old habitual ways of adapting and reacting'.[49] So doing, it may disrupt our habitual patterns, encouraging a move from reaction to reception and response. A parallel move from representation (under the hegemony of habit) to presentation (presentation that, unlike representation, may include both what is there and what is not there) enlivens everything, links us to everything – *religere* – to tie together. This opening up takes us back to the etymological roots of attention – *attendere* – to stretch out. Stretching out, we open up, touch what is other, not-self, world.

So, finally, what is of import is *practice*, a fostering of different intention and attention to all that we do that can be the only firm conclusion to take from this exploration. As Annie Dillard noted, it is a letting go, yet also a discipline that maybe requires a lifetime of dedication. Practices that sharpen attentional skills, that challenge unthinking habit, open us to new experience, instil new, better habits that in turn never remain unquestioned, are surely practices worthy of performance As neuroscience reveals, neuroplasticity is experience-dependent. The experiences we cultivate form both our selves and the perceived world. What has arisen in every conversation I have had over the course of this writing emphasizes the importance of practices of attention, of dwelling in attention, living attentively. Art and craft practice in all forms is an obvious example, mindfulness and meditation exemplify other forms, psychotherapy yet another. If I have concentrated on the experience of artists as skilled practitioners of attention, it is in the hope that, following their example, all of us may be able to consider our lives *as* works of art, projects that may, at least to some extent, be consciously created rather than continuously experienced only as reactions to outer impositions. If we can become aware of the importance of the attention we are paying at every moment, we may well become able to live better and more happily. Anything, with attention, can be a practice: gardening, walking and being aware in the world. All it takes is a little conscious attention to attention itself. William James considered that 'Attention and effort are . . . but two names for the same psychic fact.' He went on to say that one who has 'attained habits of concentrated attention, energetic volition, and self-denial in unnecessary things . . . will stand like a tower when everything rocks around him, and when his softer fellow-mortals are winnowed like chaff in the blast'.[50]

Attention is best considered not as a noun, but as a verb, a process – the process of attending. Our language of nouns unfortunately and radically entails rigidification; we habitually speak of attention rather than attending, of self rather than selfing. Rigidification excludes paradox, clings to certainty, blinkers us to the way things occur and evolve if we attend to them. Practices of attending, which are indeed practices of selfing, can be cultivated through action, through mindfulness and some effort. For process is unending, in-finite. Today's good habit, becoming habitual, fails to respond to new experience. Habit, becoming habitual, ceases to be reflective, and is no longer self-reflexive. Its restriction ceases to be noticed, yet limits the ability to respond openly to a world that is always changing. Practice of awareness must be unending.

Pay attention.

References

1 Attending to Attention

1 J. Ortega y Gasset, www.wisdomquotes.com, accessed 5 February 2015.
2 W. James, *The Principles of Psychology* (Cambridge, MA, 1981), p. 401.
3 A. Olendzki, *Unlimiting Mind* (Boston, MA, 2010), p. 85.
4 G. Watson, *A Philosophy of Emptiness* (London, 2014).
5 I am indebted to David Hinton for this description. Much of my understanding of Taoist ideas comes from his writings and translations and from the work of François Jullien.
6 See A. Alexander, 'Finding Zero: A Journey Back in Time, All for Naught', from the *New York Times*, reprinted in *The Observer* (10 May 2015).
7 *Majjhima Nikaya*, trans. Bhikkhu Nanamoli and Bhikkhu Bodhi (Somerville, MA, 1995), 19.6.
8 James, *Principles of Psychology*, Chapter IV.
9 P. Glass, *Words Without Music* (New York, 2014), p. 83.
10 O. Sacks, *On the Move* (New York, 2015) p. 368.
11 A. Olendzki, 'Shining a Light', www.tricycle.com, 28 September 2015.
12 These will be addressed in Chapter Three.
13 P. Broks, from an essay, 'Now', to be included in *Night Thoughts* (London, 2017), personal communication; N. Humphrey, *Soul Dust* (London, 2011).
14 C. Montemayor and H. Haratioun, *Consciousness, Attention and Conscious Attention* (Boston, MA, 2015).
15 *The Shorter Oxford English Dictionary* (Oxford, 1969).
16 S. Batchelor, *After Buddhism* (New York, 2015), p. 101.
17 Ibid.
18 *Dhammapada*, trans. G. Fronsdal (Boston, MA, 2011), 1.
19 Ibid., 80.
20 Ibid., 35.

21 I. McGilchrist, 'Top Brain, Bottom Brain: A Reply to Stephen
    Kosslyn & Wayne Miller', www.iainmcgilchrist.com, accessed
    2 November 2014.
22 I. McGilchrist, *The Divided Brain and the Search for Meaning*
    (ebook, nd), @ 30 per cent.
23 P. Kingsnorth 'The Witness: Opening our Eyes to the Nature
    of this Earth', www.tricycle.com, accessed 7 February 2015.
    I am indebted to him for permission to quote from this article.
24 H. Bortoft, *Taking Appearance Seriously* (Edinburgh, 2012),
    p. 191.
25 Least Heat Moon, *Blue Highways* (London, 1984), p. 17.
26 W. Berry, *Life is a Miracle* (Washington DC, 2000), pp. 138–9.
27 J. L. Adams, *Winter Music* (Middletown, CT, 2004), pp. 163–4.
28 M. Doty, *Still Life with Oysters and Lemon* (Boston, MA,
    2001), p. 68.
29 From a letter of 13 November 1925 to Witold von Hulewicz.
30 Rilke from the *Ninth Duino Elegy*.

2 The Attentive Art of Meditation and Mindfulness
Practices

1   Daito Kokushi, cited in J. Austin, *Zen and the Brain*
    (Boston, MA, 1998), p. 274.
2   While writing of Early Buddhism I have used Pali terms, for
    example Gotama.
3   S. Batchelor, *Alone With Others* (New York, 1983); *The Faith
    to Doubt* (Berkeley, CA, 1990); *Buddhism Without Beliefs*
    (New York, 1997); *Confessions of a Buddhist Atheist*
    (New York, 2010); *After Buddhism* (New Haven, CT, 2015).
4   J. Garfield, *Engaging Buddhism* (New York, 2015), p. 278.
5   *Samyutta Nikaya*, 1.22, p. 535. Where quotation arises from
    the Pali Nikayas, the collection of the *suttas* or records of the
    Buddha's talks, I have used Pali language. Otherwise and in
    relation to later theories I use Sanskrit terminology. As this is
    not an academic work, in both cases diacritics and accents are
    omitted.
6   This and following quotations arise from a personal talk with
    Stephen Batchelor, 18 July 2015.
7   E. Thompson, *Waking, Dreaming, Being* (New York, 2015),
    p. 37. This would support Garfield's description of Mahayana
    Buddhist ethics as 'moral phenomenology'.
8   P. F. Strawson, *Analysis and Metaphysics* (Oxford, 1992), p. 134.

9   J. Ganeri, *The Self* (Oxford, 2012), p. 127.
10 *The Middle Length Discourses of the Buddha*, trans. Bhikkhu Nanamoli and Bhikkhu Bhodi, p. 145.
11 L. Brasington, *Right Concentration* (Boston, MA, 2015), p. 69. I would refer readers who want more information about the *jhanas* to this volume.
12 This and the following discussion come from a private meeting with Leigh Brasington, 20 November 2015.
13 *Dzogs chen*, the Great Perfection, is a central teaching of the Tibetan *Nyingma* (Old) School of Tibetan Buddhism whose aim is the attainment and maintenance of the natural primordial state of mind (*rigpa*, see n. 14 below).
14 This refers to Mahasi Sayadaw, an influential teacher in the Burmese Theravada tradition, who taught a method of concentrative meditation that relied on noting what arises and labelling it.
15 Lama Surya Das is an American-born teacher and writer in the Tibetan *dzogs chen* tradition. *Rigpa* is a Tibetan term designating pure non-dual awareness, the natural primordial state of mind free from grasping and self-interest, see n. 12 above.
16 *Piti* and *sukha*, rapture/glee and joy/happiness, are terms used to refer to the experiences in these *jhanas*.
17 Neuroscience distinguishes between a task-orientated mode of attention and a default mode of mind wandering, which has been associated with a more self-centred perspective. This will be explained further in the following chapter.
18 Leigh Brasington, 'Unicorns Never Die', http://leighb.com/unicornsneverdie.htm, accessed 24 November 2015.
19 S.54 11 as cited in Batchelor, *After Buddhism*, p. 234.
20 Ibid.
21 See n. 6 above.
22 There are two major presentations of *Abhidharma* literature – the Theravada school *Abhidamma* (Pali) and the Sanskrit *Abhidharma*. There are differences in presentation in the two branches. I use the Sanskrit term, *Abhidharma* most generally, except when relating specifically to the Theravadin redaction.
23 For a slightly longer exposition, see G. Watson, *A Philosophy of Emptiness* (London, 2014), Chapter Three, and David Hinton, introductions to Tao-te-Ching and Chuang Tzu, inner chapters in D. Hinton, trans., *The Four Chinese Classics* (Berkeley, CA, 2013). Also see François Jullien.
24 Hinton, *The Four Chinese Classics*, p. 20.

25 Ibid., p. 137.
26 D. Hinton, trans., *Mountain Home: The Wilderness Poetry of Ancient China* (New York, 2002), p. xvii.
27 David Hinton in an interview with Leath Tonino in *The Sun*, available at www.davidhinton.net, accessed 14 October 2015.
28 Epicurus, H. Usener, ed., *Epicurea* (Leipzig, 1887), cited by Martha Nussbaum in *The Therapy of Desire* (Princeton, NJ, 1994), p. 13.
29 P. Hadot, *Philosophy as a Way of Life*, trans. A. I. Davidson (Oxford, 1995), p. 84.
30 Ibid., p. 268.
31 Ibid., pp. 208, 273.
32 Ibid., p. 59.
33 P. Sloterdijk, *You Must Change Your Life*, trans. W. Hoban (Cambridge, MA, 2013), p. 167.
34 'Modern Stoicism', www.modernstoicism.com, accessed 3 November 2016.
35 J. Kabat-Zinn, *Wherever You Go, There You Are* (New York, 2004), p. 4.
36 E. Thompson, 'The Embodied Mind', *Tricycle Magazine* (Fall 2014), p. 38.
37 Linda Heuman, 'Don't Believe the Hype', www.tricycle.com, 1 October 2014.
38 Ibid.
39 A. Olendzki, 'The Mindfulness Wedge', *Tricycle Magazine* (Fall 2014), p. 30.
40 J. Wilks, 'Secular Mindfulness', *Barre Centre for Buddhist Studies Journal* (2014).
41 Tomas Rocha, 'The Dark Knight of the Soul', www.theatlantic.com, 25 June 2014.
42 G. M. Hopkins, *Poems and Prose of Gerard Manley Hopkins*, ed. W. H. Gardner (London, 1953).
43 Willoughby Britton, 'Meditation Nation', www.tricycle.com, 25 April 2014.
44 All quotations in the following section come from a personal conversation with Jenny Wilks, 17 July 2015.
45 But we should note that Stephen Batchelor earlier had pointed out that in different presentations this is not always the case.
46 Quotations from Shinzen Young come from a conversation with him, 23 February 2016. Information about his teaching can be found at www.shinzen.org.
47 P. Iyer, *The Art of Stillness* (New York, 2015).

3 The Neuroscience of Attention

1 Sam Harris, *Waking Up* (New York, 2014).

2 E. Thompson, *Waking, Dreaming, Being* (New York, 2015), p. 172, citing H. Spiler et al., 'Increased Attention Enhances Both Behavioral and Neuronal Performance', *Science*, 15 (1988), pp. 338–40.

3 T. Elbert, C. Pantev, C. Wienbruch et al., 'Increased Cortical Representation of the Fingers of the Left Hand in String Players', *Science*, CCLXX/5234 (13 October 1995), pp. 305–7; E. A. Maguire, D. G. Gadian, I. S. Johnsrude et al., 'Navigation-related Structural Change in the Hippocampi of Taxi Drivers', *Proceedings of the National Academy of Sciences, USA*, XCVII/8 (January 2000), pp. 4398–403.

4 See N. Doidge, *The Brain That Changes Itself* (London, 2008). In this book Doidge presents a fascinating study of the discoveries and wide-ranging operations and appli-cations of neuroplasticity. Chapter Eight tells the history of the belief that thoughts can change the material structure of the brain from Thomas Hobbes in the sixteenth century through the work of philosopher Alexander Bain, Sigmund Freud and the neuroanatomist Santiago Ramón y Cajal. The specific research finding refers to K. M. Stephan, G. R. Fink, R. E. Passingham et al., 'Functional Anatomy of Mental Representation of Upper Extremity Movements in Healthy Subjects', *Journal of Neurophysiology*, LXXIII/1 (January 1995), pp. 373–86.

5 R. Hanson, *Just One Thing* (Oakland, CA, 2011), p. 1.

6 W. James, *The Principles of Psychology* (Cambridge, MA, 1981), p. 381.

7 M. Csíkszentmihályi, *Creativity: Flow and the Psychology of Discovery and Invention* (New York, 1996), p. 359; T. Metzinger, cited in Thompson, *Waking*, p. 211.

8 P. Broks, *Into the Silent Land* (London, 2003), p. 57.

9 F. Varela, E. Rosch and E. Thompson, *The Embodied Mind* (Boston, MA, 1991), p. 80.

10 Broks, *Into the Silent Land*, p. 41. In summer 2016 BBC Radio 4 aired a series of fascinating short programmes entitled *Dr Broks' Casebook*, in which Paul Broks explored the effect of various disorders on the self, which taken together show the fragility of our normal sense of self as permanent, real and objective.

11 Thompson, *Waking*, p. 346.
12 Ibid, p. 349.
13 R. Hanson, private communication, 2014.
14 M. A. Killingsworth and D. T. Gilbert, 'A Wandering Mind is an Unhappy Mind', *Science*, CCCXXX/6006 (12 November 2010), p. 932.
15 N.A.S. Farb, Z. V. Sindel, H. Mayberg et al., 'Attending to the Present: Mindfulness Meditation Reveals Distinct Neural Modes of Self-reference', *Social Cognitive Affective Neuroscience*, II/4 (December 2007), pp. 313–22.
16 Ibid., p. 314.
17 Thompson, *Waking*, pp. 348–55 and nn. 68–70.
18 P. Glass, *Words Without Music* (New York, 2014), p. 381.
19 Csíkszentmihályi, *Creativity*, pp. 345–6.
20 Ibid., p. 352.
21 Ibid., pp. 111–13.
22 I. McGilchrist, *The Divided Brain and the Search for Meaning* (ebook, nd), @ 24 per cent.
23 I. McGilchrist, 'Top Brain, Bottom Brain, A Reply to Stephen Kosslyn & Wayne Miller', www.iainmcgilchrist.com, accessed 2 November 2014.
24 This and following quotations come from a personal talk with Iain McGilchrist on Skype, 18 December 2015.
25 E. Goldberg and L. D. Costa, 'Hemispheric Differences in the Acquisition and Use of Descriptive Systems', *Brain and Language*, XIV/1 (September 1981), pp. 144–73.
26 I. McGilchrist, 'What Happened to the Soul?', www.thersa.org, 29 November 2015.
27 J-P. Lachaux, *Le Cerveau attentif* (Paris, 2011), p. 9, my translation.
28 Ibid., p. 12.
29 James, *Principles of Psychology*, p. 381.
30 S. Begley, *Train Your Mind, Change Your Brain* (New York, 2007), p. 157.
31 'Attention, Distraction and the War in Our Brain', www.youtube.com, accessed 2 April 2015.
32 Lachaux, *Cerveau*, p. 305.
33 Doidge, *Brain*, p. 68, citing M. P. Kilgard and M. M. Merzenich, 'Cortical Map Reorganization Enabled by Nucleus Basalis Activity', *Science*, CCLXXIX/5357 (March 1998), pp. 1714–18.
34 Doidge, *Brain*, p. 80.

35 Thompson, *Waking*, p. 351. A clear presentation of benefits of attentional training by Maria Konnikova, 'The Power of Concentration', appeared in an article in the *New York Times, Sunday Review*, 15 December 2012, accessed 27 December 2016.

36 J. A. Brewer et al., 'Meditation Experience is Associated with Differences in Default Network Activity and Connectivity', *Proceedings of the National Academy of Sciences, USA*, 108 (2011), pp. 20254–9, cited in Thompson, *Waking*, p. 352.

37 Ibid., p. 320, citing M. Williams, J. Teasdale and S. Zindel, *The Mindful Way Through Depression* (New York, 2007).

38 This suggestion is supported with a citation of R. Davidson, 'Well-being and Affective Style: Neural Substrates and Biobehavioural Correlates', *Philosophical Transactions of the Royal Society*, 359 (2004), pp. 1395–411.

39 Thompson, *Waking*, p. 355.

40 R. C. de Charms et al., 'Control Over Brain Activation and Pain Learned by Using Real-Time Functional MRI', *Proceedings of the National Academy of Sciences USA*, 102 (2005), pp. 1826–31, cited in Thompson, *Waking*, p. 172.

41 See www.mindandlife.org, accessed 27 December 2016.

42 R. Davidson and A. Lutz, 'Buddha's Brain: Neuroplasticity and Meditation', *IEEE Signal Processing Magazine* (September 2007), p. 176, cited in L. Brasington, *Right Concentration* (Boston, MA, 2015), p. 157.

43 S. Blackmore, *The Meme Machine* (Oxford, 2000), *Conversations on Consciousness* (Oxford, 2005), *Consciousness: A Very Short Introduction* (Oxford, 2005), *Ten Zen Questions* (Oxford, 2009), *Consciousness* (London, 2010).

44 This and the following conversation come from a private meeting with Sue Blackmore in Devon, 10 November 2015.

45 For Enactive approaches to cognitive science and philosophy of mind, see Varela et al., *Embodied Mind*. For sensori-motor approaches, see S. Gallagher, *How the Body Shapes the Mind* (Oxford, 2006), A. Noe, *Action in Perception* (Cambridge, MA, 2006), *Out of Our Heads* (New York, 2010), *Varieties of Presence* (Cambridge, MA, 2012); K. O'Regan, *Why Red Doesn't Sound Like a Bell* (Oxford, 2012). Such approaches are often referred to as 4E – Enactive, Embodied, Embedded, Extended.

46 If genes are the first replicators, then memes, carriers of cultural ideas, were the second. Sue Blackmore coined the term 'temes'

for the third and wrote of them: 'The third is that now, in the early 21st century, we are seeing the emergence of a third replicator. I call these temes (short for technological memes, though I have considered other names). They are digital information stored, copied, varied and selected by machines. We humans like to think we are the designers, creators and controllers of this newly emerging world but really we are stepping stones from one replicator to the next.' S. Blackmore, 'The Third Replicator', http://opinionator.blogs.nytimes.com, 20 August 2010.

47 Doidge, *Brain*, p. 309.
48 'SSHRC Insight Grant', www.evanthompson.com, 13 April 2013.
49 The following comes from a private interview with Rick Hanson, 16 March 2015.
50 Nonoguchi Ryuho (1595–1669).
51 See www.rickhanson.net/writings/just-one-thing.

4 Emotional Attention

1 M. Oliver cited in M. Popova, 'Mary Oliver on What Attention Really Means', www.brainpickings.org, accessed 28 March 2016.
2 S. Gerhardt, *Why Love Matters* (London, 2004).
3 D. Goleman, www.wisdomquotes.com, accessed 5 February 2015.
4 Gerhardt, *Love*, p. 9.
5 Ibid., referring to A. N. Schore, *Affect Regulation and the Origin of the Self* (Hillsdale, NJ, 1994).
6 A. N. Schore, 'The Effects of Early Relational Trauma on Right Brain Development, Affect Regulation, and Infant Mental Health', *Infant Mental Health Journal*, XXII (2001), pp. 201–69.
7 A. N. Schore, 'The Effects of a Secure Attachment Relationship on Right Brain Development, Affect Regulation and Infant Mental Health', *Infant Mental Health Journal*, XXII/201–60 (2001), pp. 227–66.
8 Ibid.
9 D. Zahavi and P. Rochat, 'Empathy≠Sharing: Perspectives from Phenomenology and Developmental Psychology', in *Consciousness and Cognition* (nd), www.academia.eu, accessed 27 January 2015.
10 Schore, 'The Effects of Early Relational Trauma', p. 42.

11 A. N. Schore, *Neuro-Psychoanalysis*, I. 115–128, p. 125.
12 A. N. Schore, *Affect Dysregulation* and *Disorders of the Self and Affect Regulation and the Repair of the Self*, vol. II (New York, 2003).
13 What follows comes from a conversation with Maura Sills, 5 August 2015.
14 C. Trungpa, *The Sanity We are Born With*, ed. C. R. Gimian (Boston, MA, 2005), cited in K. K. Wegela, *Contemplative Psychotherapy Essentials* (New York, 2014), p. 5.
15 Wegela, ibid., p. 16.
16 Ibid, p. 158.
17 The following dialogue arises from a Skype talk with Karen Kissel Wegela, 8 October 2015.
18 Buddhist teachings are often distinguished as three types: Early Buddhism, pejoratively called Hinayana; later Mahayana; and Vajrayana, the tantric-influenced teachings of Tibet.
19 From a personal communication with Shinzen Young, 23 February 2016.

5 Attentive Education

1 W. James, *The Principles of Psychology* (Cambridge, MA, 1981), p. 401.
2 M. Crawford, *Shop Class as Soulcraft* (New York, 2009); M. Crawford, *The World Beyond Your Head* (New York, 2015); 'Hey You', Matthew Crawford talking with Ed Cumming in *The Guardian* (11 April 2015).
3 From Rosie Ifould, 'Screen Break', in *Guardian Weekend* (18 June 2016).
4 Crawford, *World*, p. 251.
5 Ibid., p.16.
6 Ibid., p. x.
7 Ibid., p. 121.
8 Ibid., p. 117.
9 R. Allen, *Polanyi* (London, 1990), p. 35.
10 Ibid, p. 37.
11 Crawford, *World*, p. 24.
12 Ibid., p. 23.
13 Ibid.
14 Ibid., pp. 23–4.
15 Ibid.

16 Allen, *Polanyi*, p. 51.
17 Ibid., p. 39.
18 Crawford, *World*, p. 25.
19 Crawford, *Shop Class*, p. 82.
20 M. Heidegger, *What is Called Thinking?*, trans. J. Glenn Gray (New York, 1976), p. 15.
21 Ibid., pp. 16–17.
22 A. Clark, *Being There: Putting Brain, Body and World Together Again* (Cambridge, MA, 1997) and *Supersizing the Mind* (New York, 2008); A. Noe, *Out of Our Heads* (New York, 2010), *Varieties of Presence* (Cambridge, MA, 2012), *Strange Tools, Art and Human Nature* (New York, 2015); J. K. O'Regan, *Why Red Doesn't Sound Like a Bell* (Oxford, 2012); S. Gallagher, *How the Body Shapes the Mind* (Oxford, 2006), *Phenomenology* (2012); G. Lakoff and M. Johnston, *Philosophy in the Flesh* (New York, 1999). Also see Chapter Three, n. 45.
23 Crawford, *World*, p. 68.
24 Ibid., p. 257.
25 Ibid., p. 253.
26 Ibid., p. 257.
27 The following quotations from Professor Claxton come from a meeting in London, 27 April 2015.
28 J. L. Roberts, 'The Power of Patience', *Harvard Magazine* (November/December 2013).
29 K. Weare, 'Developing Mindfulness with Children and Young People: A Review of the Evidence and Policy Context', *Journal of Children's Services*, VIII/2 (March 2013), pp. 141–53, and www.mindfulnessinschools.org, accessed 15 July 2015.
30 Ibid.
31 All information about these programmes and trainings can be found at www.mindfulnessinschools.org.
32 Weare, 'Developing Mindfulness' gives a presentation of much of the evidence to date.
33 M. Bunting, 'Why We Will Come to See Mindfulness as Mandatory', www.theguardian.com, 6 May 2014.

Part Two: Attending Creatively

1 L. Anderson, interview with T. Teeman, www.thedailybeast.
   com, 29 September 2015, and www.believermag.com, January
   2012, accessed 22 February 2016.
2 G. Steiner, *Real Presences* (London, 1989), p. 202.
3 Ibid., p. 209.
4 H. Shukman, 'The Unfamiliar Familiar', www.tricycle.com,
   14 January 2014.
5 Ibid.
6 Guy de Maupassant, *Une Vie* (Paris, 1883), p. 22, cited in
   H. Spurling, *The Unknown Matisse* (London, 2000), p. 167.
7 J. Hirshfield, *Ten Windows* (New York, 2015), p. 12.
8 J. Hirshfield, 'Felt in its Fullness', from M. Matousek,
   'An Interview with Jane Hirshfield', www.tricycle.com,
   11 April 2014.
9 Hirshfield, *Ten Windows*, p. 177.
10 Ibid., pp. 12, 35.
11 Ibid, p.182.
12 Hirshfield, 'Felt in its Fullness'.
13 B. C. Lane, *Backpacking with the Saints: Wilderness Hiking
   as a Spiritual Practice* (Oxford, 2015), p. 98.

6 Attentiveness to the Word

1 S. Sontag, from a speech upon being awarded the
   Friedenspreis des Deutschen Buchhandels (Peace Prize of the
   German Book Trade), Frankfurt Book Fair, 12 October 2003.
2 M. Proust, *Time Regained*, trans. C. K. Scott Moncrieff and
   T. Kilmartin, rev. D. Enright (New York, 1992), pp. 299–300.
3 M. Heidegger, *Poetry, Language Thought*, trans. A. Hofstadter
   (New York, 1971), p. 215.
4 M. Heidegger, *What is Called Thinking?*, trans. J. Glenn Gray
   (New York, 1976), p. 14.
5 M. Heidegger, *On the Way to Language*, trans. P. D. Hertz
   (New York, 1971), p. 57.
6 Ibid., p. 123.
7 Ibid, p. 134.
8 L. Hogan, www.wisdomquotes.com, accessed
   5 February 2015.
9 H. Miller, ibid.

10 W. Berry, *Life is a Miracle* (Washington, DC, 2000), p. 138.
11 Ibid.
12 R. MacFarlane, *The Wild Places* (London, 2007), p. 225.
   Deakin's books are *Waterlog* (London, 2003), *Wildwood*
   (London, 2007) and *Notes from Walnut Tree Farm* (London,
   2008).
13 MacFarlane, *Wild Places*, p. 321.
14 R. MacFarlane, *Landmarks* (London, 2015), p. 63.
15 R. MacFarlane, 'The Ultimate Wilderness', *The Guardian*
   (11 July 2015).
16 MacFarlane, *Landmarks*, p. 240.
17 Ibid., p. 35.
18 N. Shepherd, *The Living Mountain* (Edinburgh, 2011),
   pp. 10–11.
19 The following arises from a talk with Ruth Ozeki,
   30 April 2016, in Barre, MA.
20 Cited in J. Hirshfield, *Ten Windows* (New York, 2015),
   p. 296.
21 This and following quotations come from a meeting with
   Jane Hirshfield in Marin, CA, 27 February 2016.
22 Hirshfield, *Ten Windows*, p. 7.
23 Ibid., p. 274.
24 J. Keats in a letter to his brothers George and Thomas Keats,
   22 December 1817.
25 Hirshfield, *Ten Windows*, p. 122.
26 Izumi Shikibu, trans. J. Hirshfield and M. Aratani, as quoted
   in Hirshfield, *Ten Windows*, p. 124.
27 Ibid., p. 125.
28 Ibid., p. 52.
29 M. Gonnerman, ed., *A Sense of the Whole* (Berkeley,
   CA, 2015), p. 264.
30 Hirshfield, *Ten Windows*, p. 90.
31 Ibid., p. 252.
32 J. Felstiner, *Can Poetry Save the World?* (New Haven,
   CT, 2009), p. 155.
33 Heidegger, *Thinking*, p. 128.
34 M. Heidegger, 'The Origin of the Work of Art', in *Poetry*,
   pp. 74, 75, 72.
35 Ibid., p. 12.
36 Ibid., p. 75.
37 D. Hinton, *The Four Chinese Classics* (Berkeley, CA, 2013),
   p. 20.

38 D. Hinton, *Hunger Mountain* (Boulder, CO, 2012), p. 30.
39 This and following quotations come from a personal conversation with David Hinton, 15 February 2016.
40 D. Hinton, *Existence* (Boulder, CO, 2016).
41 O. Paz, 'The Other Voice', cited in T. Dean 'The Other's Voice', in Gonnerman, *Sense of the Whole*, p. 47.
42 W. Bronk, 'Ergo Non Sum, Est', in *Life Supports* (Northfield, MA, 1997), p. 147.
43 W. Bronk, 'Bare Boards at the Globe', in *Bursts of Light* (Northfield, MA, 2012), p. 139.
44 W. Bronk, 'The Tree in the Middle of the Field', in *Life Supports*, p. 49.
45 W. Bronk, *The Lens of Poetry*, ed. W. S. Hurst (New Rochelle, NY, 2011), np.
46 For this fact and an illuminating piece about Bronk's work, I am indebted to Daniel Woolf, 'Why Nobody Reads William Bronk', www.theliteraryreview.org, accessed 2 February 2015.
47 J. Gilbert, *Collected Poems* (New York, 2012); from *Winter in the Night Fields*, *The Answer* and *Ovid*.
48 P. Celan, *Selected Poetry and Prose*, trans. J. Felstiner (New York, 2001), p. 396.
49 Ibid., pp. 406–10.
50 From a review of *Falling Awake* by Fiona Sampson, *Guardian*, 19 August 2016.
51 This and the following quotations from Alice and Peter Oswald come from a meeting with them in Devon, 7 May 2015.
52 In an interview with Max Porter in *The White Review*, www.thewhitereview.org, August 2014.
53 In an interview with Stephen Knight, www.independent.co.uk, 5 April 2009.
54 Michael Chekhov (1891–1955) was an influential actor, director and theatre practitioner.

7 Visual Attention

1 A. Glimcher, ed., *Agnes Martin, Painting, Writings, Remembrances* (London and New York, 2012), p. 11.
2 A. Ehrenzweig, *The Hidden Order of Art* (Berkeley, CA, 1967), p. 25.
3 From a descriptive note to an Ad Reinhardt painting in the Anderson Collection at Stanford University (27 February 2016).

4 A. Gormley, *On Sculpture* (London, 2015), p. 36.

5 Ibid., p. 41.

6 Ibid., p. 12.

7 From a descriptive note to Sean Scully work in the Anderson Collection, see n. 3 above.

8 *Take Your Time*. A travelling exhibition of the work of Olafur Eliasson, 2008–9, shown at San Francisco Museum of Modern Art, April–June 2008. See www.olafureliasson.net.

9 J. Turrell, *Deer Shelter*, exh. cat., Yorkshire Sculpture Park (Wakefield, 2006).

10 Quoted by A. Graham-Dixon in *James Turrell: A Life in Light* (London, 2007), p. 37.

11 Ibid.

12 M. Govan, 'Inner Light', in M. Govan and C. Y. Kim, *James Turrell: A Retrospective*, exh. cat., Los Angeles County Museum of Art (2012), p. 34.

13 C. Kim, 'Entering the New Landscape', in Govan and Kim, *James Turrell*, p. 250, citing both Roberta Smith and Markus Brüderlin.

14 Miwon Kwon, 'Rooms for Light, Light On its Own', in *James Turrell*, exh. cat., Gagosian Gallery (London, nd).

15 J. L. Adams, *Winter Music* (Middletown, CT, 2004), p. 164.

16 A. Ross, *And the Rest is Noise* (New York, 2007), p. 214.

17 G. Didi-Huberman, 'The Fable of the Place', in *Turrell: The Other Horizon*, ed. D. Birnbaum et al. (Berlin, 1999), p. 48.

18 J. Turrell, *Sensing Space, Seattle, Henry Art Gallery 1992*, cited in Govan and Kim, *James Turrell*, p. 212.

19 M. Kwon, 'Rooms'.

20 Quotations from James Turrell in this section come from a personal conversation at Tremenheere Gardens, Penzance, 7 June 2015.

21 A koan is the often-paradoxical question (the best known perhaps being 'What is the sound of one hand clapping?') that a Zen master addressed to a student to meditate upon. Its aim is to defamiliarize our normal habits of dualistic conception and point to what lies outside that framework.

22 M. Barnes, introduction to Garry Fabian Miller, *Home Dartmoor* (Exeter, 2012), p. 20.

23 Ibid.

24 J. Ortega y Gasset, www.wisdomquotes.com, accessed 5 February 2015.

25 In autumn 2016 Fabian Miller curated an exhibition at the
Crafts Study Centre, Farnham, including work by Batterham
and other craftsmen. The exhibition, titled 'Making Thinking
Living', illustrated and celebrated the integration of a way of
life, making and thought.

26 Barnes, introduction, p. 20.

27 Ibid., p. 60, quoting from Virginia Woolf's diary for
26 February 1926.

28 V. Woolf, 'A Sketch of the Past', in *Moments of Being*
(London, 2002).

29 S. Batchelor, 'Seeing the Light: Photography as Buddhist
Practice', in *Buddha Mind in Contemporary Art*, ed. J. Baas
and M. J. Jacob (Berkeley, CA, 2004), p. 141.

30 From a personal conversation with Stephen Batchelor at
Gaia House, Devon, 18 July 2015.

31 See www.middlewaysociety.org, podcast episode 7
(December 2013).

32 Ibid.

33 S. Batchelor, 'Aesthetics of Emptiness', a talk given at BCBS,
Barre, on 28 April 2016, to be published in 2017, in S.
Batchelor, *Secular Buddhism: Imagining the Dharma in an
Uncertain World* (New Haven, CT, 2017).

34 Edmund de Waal, 'White', RA *Magazine* (autumn 2015), p. 70.
Unless otherwise stated, quotation from de Waal below comes
from a conversation at his studio in London, 3 September
2015.

35 De Waal, *Atemwende*, exh. cat., Gagosian Gallery (New York,
2013).

36 E. de Waal, *The White Road* (London, 2015).

37 S. Anderson, 'Edmund de Waal, the Strange Alchemy of
Porcelain', *New York Times* (25 November 2015).

38 F. Hallett, review of Edmund de Waal, 'I Placed a Jar'
exhibition at Brighton Festival, www.theartsdesk.com,
19 May 2015.

39 See the following chapter for reference to John Cage, 4'33".

40 The following quotations come from a conversation with
Pablo Bronstein, 30 November 2015. Bronstein's 'Historical
Dancers in an Antique Setting' provided the summer
installation at the Duveen Gallery of Tate Britain in 2016. In
November 2016, he designed the costumes and sets for Ballet
Rambert's new dance staging of Joseph Haydn's *The Creation*.

8 Aural Attention: Listening and Hearing

1 John Luther Adams, 'Making Music in the Anthropocene', *Slate Magazine* (24 February 2015), and *Winter Music* (Middletown, CT, 2004), p. 164.

2 G. Steiner, *Errata* (London, 1997), p. 65.

3 Ibid, p. 67.

4 A. Huxley, *Music at Night* (London, 1931), p. 19.

5 Adams, *Winter Music*, p. 111.

6 P. Glass, *Words Without Music* (New York, 2015), p. 95.

7 A. Brendel, *Music, Sense and Nonsense* (London, 2015), p. 324.

8 An out-of-court settlement was made.

9 K. Larson, *Where the Heart Beats* (New York, 2012), pp. 275–7, quoting from R. Fleming and W. Duckworth, eds, *John Cage at Seventy-Five* (Cranbury, NJ, 1989), pp. 21–2.

10 Glass, *Words*, p. 96.

11 Adams, *Winter Music*, p. 79.

12 Tom Service, 'The Musical and Environmental Mindfulness of John Luther Adams', www.theguardian.com, 26 February 2015.

13 Adams, 'Making Music in the Anthropocene'.

14 A. Ross, 'The Song of the Earth', www.newyorker.com, 12 May 2008.

15 Alexandra Coghlan, 'Lost in Thought', www.theartsdesk.com, 25 September 2015.

16 See www.johnlutheradams.net, accessed 22 January 2016.

17 This, Turrell's first public sky space (1986), has been closed for the past few years, but reopened in summer 2016 after renovation. PS1, now associated with the Museum of Modern Art, is an organization devoted to the advancement of contemporary art.

18 Tom Service, 'Music Matters', BBC Radio 3 with Kevin Volans, Richard Bonas and John Tilbury (2 October 2015). The following quotations come from this audio programme.

19 C. Rothko, *From the Inside Out* (New Haven, CT, 2015), p. 120.

20 Ibid., pp. 126–7.

21 Ibid., p. 127.

22 Ibid., p. 180.

23 From programme notes by James M. Keller for a performance conducted by Michael Tilson Thomas with the San Francisco Symphony, 23 February 2011. I am grateful to Mr Keller for permission to quote from these notes, which have also appeared in G. Watson, *A Philosophy of Emptiness* (London, 2014), p. 157.

24 P. Oliveros, *Deep Listening: A Composer's Sound Practice* (New York, 2005), pp. xvii–xxv.

25 The following quotations come from a meeting with Sam Richards and Lona Kozik in Devon, 3 August 2015.

9 Embodied Attention

1 F. Nietzsche, *Thus Spake Zarathustra*, www.lexido.com, accessed 18 October 2016.

2 N. Doidge, *The Brain's Way of Healing* (London, 2015), p. 169.

3 Adapted from ibid., pp. 169–76.

4 All quotations in this section come from a personal conversation with Catherine McCrum, 17 July 2015.

5 Flow is a term from the work of Mihály Csíkszentmihályi (1990), which describes optimum experiential moments of flow as being those when the balance between the difficulty or danger of the task is matched by the ability and attention of the practitioner. In such moments, self-consciousness and wandering mind are absent. See Chapter Three.

6 This and the following quotations come from a Skype conversation with Wayne McGregor, 16 December 2015.

7 This and the following quotations come from a conversation with Robert Silliman, 19 August 2015.

8 S. Bleakley, *Mindfulness and Surfing* (Lewes, 2016), p. 9.

9 Ibid., p. 4.

10 Ibid., p. 17.

11 Ibid., p. 81.

12 Subsequent quotations come from a Skype conversation with Sam Bleakley, 27 November 2015.

13 W. Finnegan, *Barbarian Days* (New York, 2015); D. Duane, *Caught Inside* (New York, 1996).

14 Duane, *Caught Inside*, pp. 219 and 235.

15 Bleakley, *Surfing*, p. 115.

10 Attentive and Experiential In-conclusions

1 A maxim of Malebranche beloved by Simone Weil and Paul Celan.
2 Dale S. Wright, 'Religion Resurrected', www.tricycle.com (3 August 2015). An extended version of this argument appears in Chapter 3 of D. S. Wright, *What is Buddhist Enlightenment?* (Oxford, 2016).
3 Jungu Yoon, *Spirituality in Contemporary Art* (London, 2015).
4 In a letter to his brothers George and Thomas Keats, 2 December 1817.
5 T. S. Eliot, *Dry Salvages*, in *Four Quartets* (London, 1943).
6 J. Campbell, *The Power of Myth* (New York, 1988), p. 5.
7 S. Beckett, 'Texts for Nothing 4', in *The Complete Short Prose* (New York, 1995), p. 116.
8 D. Hinton, *Hunger Mountain* (Boston, MA, 2012), p. 124.
9 R. Porter, *Flesh in the Age of Reason* (London and New York, 2003), p. 181.
10 Cited in A. Harris, *Weatherland* (London, 2015), pp. 251–2.
11 F. Varela, E. Rosch and E. Thompson, *Embodied Mind*, p. 80.
12 A. Dillard, *Pilgrim at Tinker Creek* (New York, 1974), p. 258.
13 Ibid., p. 97.
14 Batchelor, *Confessions of a Buddhist Atheist* (New York, 2010), pp. 127–8. See also *After Buddhism*, pp. 55–6.
15 A. Gormley, *On Sculpture* (London, 2015), p. 143.
16 Ibid., p. 172.
17 G. Claxton, *Intelligence in the Flesh* (New Haven, CT, 2015), p. 242.
18 I. McGilchrist, *The Master and His Emissary* (New Haven, CT, 2009), p.133.
19 B. Lopez, 'The Invitation', www.granta.com (November 2015).
20 See Chapter One, n. 20.
21 H. Bortoft, *Taking Appearance Seriously* (Edinburgh, 2012), p. 25. I am indebted to Bortoft's presentation here, and to discussion about this with my great discussion group in Point Reyes.
22 M. Heidegger, *Introduction to Metaphysics* (New Haven, CT, 2000), p. 107, and as cited in Bortoft, *Appearance*, p. 25.
23 M. Heidegger, *On the Way to Language* (New York, 1971), p. 123, cited in Bortoft, *Appearance*, p. 135.
24 Ibid, p. 134.

25 P. Cézanne, quoted in B. Dorival, *Paul Cézanne*, trans. H.H.A. Thackthwaite (London, 1948), cited in *The Merleau-Ponty Aesthetics Reader*, ed. G. A. Johnson (Evanston, IL, 1993).

26 L. Berger, 'Being There: Heidegger on Why Our Presence Matters', www.newyorktimes.com (30 March, 2015).

27 Ibid.

28 Lopez, 'The Invitation'.

29 Emily Dickinson, 'Forever – is Composed of Nows', www. poetryfoundation.org, accessed 23 May 2016.

30 S. Batchelor, *After Buddhism* (New Haven, CT, 2015), pp. 231.

31 Ibid.

32 Ibid.

33 I. McGilchrist, 'What Happened to the Soul?', www.thersa. org, 31 March 2014.

34 Podcast www.middlewaysociety.org, 15 February 2015.

35 Batchelor, *After Buddhism*, p. 340.

36 See n. 1.

37 W. H. Auden, 'The Nature of Prayer', in *A Certain World* (New York, 1970).

38 K. Jamie, *Findings* (London, 2005), p. 109.

39 T. Deacon, *Incomplete Nature* (New York, 2013), p. 534.

40 Ibid., Glossary.

41 Ibid., p.540.

42 Ibid., Glossary.

43 Ibid., pp. 544 and 541.

44 I. McGilchrist, 'The Power of No', Laing Lecture 3, Regent College, Vancouver, 2016.

45 I. McGilchrist, 'What Brains Can and Can't Teach us About God', Laing Lecture 2, Regent College, Vancouver, 2016, and 'The Power of No'.

46 S. Kierkegaard, 'The Absolute Paradox: A Metaphysical Caprice', *Philosophical Fragments*, trans. H. V. Hong and E. H. Hong (Princeton, NJ, 1985) p. 37, cited in McGilchrist, 'What Brains'.

47 Dillard, *Pilgrim*, pp. 31, 32. I am indebted to Maria Popova, 'Annie Dillard on How to Live with Mystery, the Two Ways of Looking and the Secret of Seeing', in www.brainpickings.org, accessed 28 March 2016, for reminding me of this.

48 Claxton, *Intelligence*, p. 250.

49 D. Siegel, *The Mindful Brain* (New York, 2007), p. 127.

50 W. James, *The Principles of Psychology* (Cambridge, MA, 1981), p. 130.

Bibliography

Adams, J. L., *Winter Music* (Middletown, CT, 2004)
Allen, R., *Polanyi* (London, 1990)
Auden, W. H., *A Certain World* (London, 1982)
Austin, J., *Meditating Selflessly, Practising Neural Zen*
    (Boston, MA, 2011)
—, *Zen and the Brain* (Boston, MA, 1998)
—, *Zen Brain Horizons* (Boston, MA, 2014)
Baas, J., and M. J. Jacob, eds, *Buddha Mind in Contemporary Art*
    (Berkeley, CA, 2004)
Baker, I., *The Heart of the World* (New York, 2004)
Batchelor, S., *After Buddhism* (New Haven, CT, 2015)
—, *Alone With Others* (New York, 1983)
—, *Buddhism Without Beliefs* (New York, 1997)
—, *Confessions of a Buddhist Atheist* (New York, 2010)
—, *The Faith to Doubt* (Berkeley, CA, 1990)
—, *Secular Buddhism: Imagining the Dharma in an Uncertain*
    World (New Haven, CT, 2017)
Beckett, S., *The Complete Short Prose* (New York, 1995)
Begley, S., *Train Your Mind, Change Your Brain*
    (New York, 2007)
Berry, W., *Life is a Miracle* (Washington, DC, 2000)
Birnbaum, D., *James Turrell, The Other Horizon*
    (Vienna, 1999)
Blackmore, S., *Consciousness: A Very Short Introduction*
    (Oxford, 2005)
—, *Consciousness* (London, 2010)
—, *The Meme Machine* (Oxford, 2000)
—, *Ten Zen Questions* (Oxford, 2009)
Bleakley, S., *Mindfulness and Surfing* (Lewes, 2016)
Bortoft, H., *Taking Appearance Seriously* (Edinburgh, 2012)
Brasington, L., *Right Concentration* (Boston, MA, 2015)
Brendel, A., *Music, Sense and Nonsense* (London, 2015)
Broks, P., *Into the Silent Land* (London, 2003)
—, *Night Thoughts* (London, 2017)

Bronk, W., *Bursts of Light* (Northfield, MA, 2012)
—, *The Lens of Poetry* (New Rochelle, NY, 2012)
—, *Life Supports* (Northfield, MA, 1997)
Campbell, J., *The Power of Myth* (New York, 1998)
Celan, P., *Selected Poetry and Prose*, trans. J. Felstiner
    (New York, 2001)
Claxton, G., *Intelligence in the Flesh* (New Haven and
    London, 2015)
Crawford, M., *Shop Class as Soulcraft* (New York, 2009)
—, *The World Beyond Your Head* (New York, 2015)
Csíkszentmihályi, M., *Creativity: Flow and the Psychology
    of Discovery and Invention* (New York, 1996)
Deacon, T., *Incomplete Nature* (New York, 2013)
Deakin, R., *Notes from Walnut Tree Farm* (London, 2008)
—, *Waterlog* (London, 2003)
—, *Wildwood* (London, 2007)
Dhammapada, trans. G. Fronsdal (Boston, MA, 2011)
Dillard, A., *Pilgrim at Tinker Creek* (New York, 1974)
Doidge, N., *The Brain That Changes Itself* (London, 2008)
—, *The Brain's Way of Healing* (London, 2015)
Doty, M., *Still Life With Oysters and Lemon* (Boston, MA, 2001)
Duane, D., *Caught Inside* (New York, 1996)
Edelman, G., *Bright Air, Brilliant Fire* (New York, 1992)
—, *Neural Darwinism* (New York, 1987)
Ehrenzweig, A., *The Hidden Order of Art* (Berkeley, CA, 1967)
Eliot, T. S., *Collected Poems* (London, 1963)
Felstiner, J., *Can Poetry Save the World?* (New Haven, CT, 2009)
Finnegan, W., *Barbarian Days* (New York, 2015)
Gallagher, W., *Rapt* (New York, 2009)
Ganeri, J., *The Self* (Oxford, 2012)
Gerhardt, S., *Why Love Matters* (London, 2004)
Gilbert, J., *Collected Poems* (New York, 2012)
Glass, P., *Words Without Music* (New York, 2014)
Gonnerman, M., ed., *A Sense of the Whole* (Berkeley, CA, 2015)
Gormley, A., *On Sculpture* (London, 2015)
Graham-Dixon, A., *James Turrell, A Life in Light*
    (London, 2007)
Hadot. P., *Philosophy as a Way of Life*, trans. A. I. Davidson
    (Oxford, 1995)
Hanson, R., *Buddha's Brain* (Oakland, CA, 2009)
—, *Hardwiring Happiness* (New York, 2013)
—, *Just One Thing* (Oakland, CA, 2011)

Harris, S., *Waking Up* (New York, 2014)

Heat Moon, W. L., *Blue Highways* (New York, 1982)

Heidegger, M., *Introduction to Metaphysics*, trans. G. Fried
    and R. Polt (New Haven, CT, 2000)

—, *On the Way to Language*, trans. P. D. Hertz (New York, 1971)

—, *Poetry, Language, Thought*, trans. A. Hofstadter
    (New York, 1971)

—, *What is Called Thinking?*, trans. J. Glenn Gray (New York,
    1976)

Hinton, D., trans., *The Four Chinese Classics* (Berkeley,
    CA, 2013)

—, trans., *Hunger Mountain* (Boulder, CO, 2016)

—, trans., *Mountain Home: The Wilderness Poetry of Ancient
    China* (New York, 2002)

Hirshfield, J., *Nine Gates* (New York, 1997)

—, *Ten Windows* (New York, 2015)

Hirshfield, J., and M. Aratani, trans., *The Ink Dark Moon*
    (New York, 1990)

Horowitz, A., *On Looking* (New York, 2013)

Huxley, A., *Island* (New York, 1962)

—, *Music at Night* (London, 1931)

Iyer, P., *The Art of Stillness* (New York, 2015)

James, C., *Cultural Amnesia: Notes in the Margins of my Time*
    (London, 2016)

James, W., *The Principles of Psychology* (Cambridge, MA, 1981)

Kabat-Zinn, J., *Wherever You Go, There You Are* (New York,
    2004)

Lachaux, J-P., *Le Cerveau attentif* (Paris, 2011)

Lane, B. C., *Backpacking with the Saints: Wilderness Hiking
    as a Spiritual Practice* (Oxford, 2015)

Larson, K., *Where the Heart Beats* (New York, 2012)

McEvilley, T., *The Shape of Ancient Thought* (New York, 2002)

MacFarlane, R., *Landmarks* (London, 2015)

—, *The Old Ways* (London, 2012)

—, *The Wild Places* (London, 2007)

McGilchrist, I., *The Divided Brain and the Search for Meaning*
    (ebook, nd)

—, *The Master and His Emissary* (New Haven, CT, 2009)

Nussbaum, M., *The Therapy of Desire* (Princeton, NJ, 1994)

Olendzki, A., *Unlimiting Mind* (Boston, MA, 2010)

Oliveros, P., *Deep Listening: A Composer's Sound Practice*
    (New York, 2005)

Oswald, A., *Dart* (London, 2002)
—, *Falling Awake* (London, 2016)
—, *Memorium* (London, 2012)
Oswald, P., and S. Borodale, *Dyad* (2005)
Ross, A., *And the Rest is Noise* (New York, 2007)
Rothko, C., *From the Inside Out* (New Haven, CT, 2015)
Sacks, O., *On the Move* (New York, 2015)
Samyutta Nikaya, trans. Bikkhu Bodhi (Boston, MA, 2000)
Schore, A. N., *Affect Dysregulation and Disorders of the Self* and *Affect Regulation and the Repair of the Self*, 2 vols (New York, 2003)
—, *Affect Regulation and the Origin of the Self* (1994)
—, 'The Effects of a Secure Attachment Relationship on Right Brain Development', *Infant Mental Health Journal*, XXII/201-160, reprinted at www.trauma-pages.com
Shepherd, N., *The Living Mountain* (Edinburgh, 2011)
Siegel, D., *The Mindful Brain* (New York, 2007)
Sloterdijk, P., *You Must Change Your Life*, trans. W. Hoban (Cambridge, MA, 2013)
Spurling, H., *The Unknown Matisse* (London, 2000)
Steiner, G., *Errata* (London, 1997)
—, *Real Presences* (London, 1989)
Thompson, E., *Waking, Dreaming, Being* (New York, 2015)
Varela, F., E. Rosch and E. Thompson, *The Embodied Mind* (Boston, MA, 1991)
Waal, E. de, *The White Road* (London, 2015)
Watson, G., *A Philosophy of Emptiness* (London, 2014)
Wegela, K. K., *Contemplative Psychotherapy Essentials* (New York, 2014)
Williams, M., J. Teasdale and S. Zindel, *The Mindful Way Through Depression* (New York, 2007)
Wittgenstein, L., *Tractatucs Logico-Philosophicus*, trans. D. F. Pears and B. F. McGuinness (London, 1960)
Woolf, V., *Moments of Being* (London, 2002)
Wright, D. B., *What is Buddhist Enlightenment?* (Oxford, 2016)
Yoon, J., *Spirituality in Contemporary Art* (London, 2015)

Acknowledgements

More than anything I have ever written before, this book has been collaborative. I am indebted to so many people who have generously given me their time and knowledge to explore the subject of attention.

My first gratitude goes to those I interviewed specifically for the text; all of those with whom I have spoken directly and at length, who provided me with so much material and such wonderful meetings along the path:

John Luther Adams, Stephen Batchelor, Susan Blackmore, Sam Bleakley, Leigh Brassington, Pablo Bronstein, Guy Claxton, Edmund de Waal, Rick Hanson, David Hinton, Jane Hirshfield, Lona Kozik, Catherine McCrum, Ian McGilchrist, Wayne McGregor CBE, Garry Fabian Miller, Alice and Peter Oswald, Ruth Ozeki, Sam Richards, Robert Silliman, Maura Sills, James Turrell, Karen Kissel WegelaJ and Jenny Wilks.

Then there were others with whom I spoke more informally, to whom I am indebted for ideas, quotation and inspiration, Paul Broks, Madeleine Bunting, Karen Houston, Paul Kingsnorth, Evan Thompson, Kathleen Weare, Dale S. Wright and Shinzen Young. I owe much also to Neil Armstrong of Tremenheere Gardens in Cornwall for an introduction to James Turrell and also for the inspiration of an amazing place.

Beyond this group there lies another circle of friends who are due my thanks, having been in less obvious ways, no less materially helpful over the time of gathering and writing. In Devon I thank my friends, Bridget McCrum, Francis Gynn, Tina and Paul Riley, Sally and Paul Vincent, Tinker and Douglas Stoddart, for their suggestions, support and much good talking and walking. In California my gratitude goes to my wonderful discussion group in Point Reyes, John Gouldthorpe, Art Levit and, for some of the time, Patricia Berry.

Fourthly, it has been a pleasure working, now on a second book, with a supportive editor Ben Hayes, and also with Martha Jay,

Howard Trent, Maria Kilcoyne and Simon McFadden at Reaktion. Extra thanks are also due to Garry Fabian Miller who has allowed me to use his wonderful image Gaze, which is such an apposite and beautiful cover.

And as ever, special thanks go to Stephen Batchelor who has been, as always a source of great discussion and fertile ideas over good dinners. Finally, my gratitude goes to my family, David, Ash, Rowan, Maryam and James, not to mention my grandchildren Johan and Ayla, who are currently teaching me a lot about developing attention theirs and mine.